Differential Operators on
**Spaces of Variable
Integrability**

Differential Operators on
Spaces of Variable
Integrability

David E. Edmunds
University of Sussex, UK

Jan Lang
The Ohio State University, USA

Osvaldo Méndez
The University of Texas at El Paso, USA

 **World Scientific**

NEW JERSEY · LONDON · SINGAPORE · BEIJING · SHANGHAI · HONG KONG · TAIPEI · CHENNAI

Published by

World Scientific Publishing Co. Pte. Ltd.

5 Toh Tuck Link, Singapore 596224

USA office: 27 Warren Street, Suite 401-402, Hackensack, NJ 07601

UK office: 57 Shelton Street, Covent Garden, London WC2H 9HE

Library of Congress Cataloging-in-Publication Data

Edmunds, D. E. (David Eric)

Differential operators on spaces of variable integrability / David E. Edmunds (University of Sussex, UK), Jan Lang (The Ohio State University, USA), Osvaldo Mendez (The University of Texas at El Paso, USA).

pages cm

Includes bibliographical references and indexes.

ISBN 978-9814596312 (hardcover : alk. paper)

1. Function spaces. 2. Sobolev spaces. 3. Differential operators. I. Lang, Jan. II. Mendez, Osvaldo (Osvaldo David) III. Title.

QA323.E25 2014

515'.73--dc23

2014015285

British Library Cataloguing-in-Publication Data

A catalogue record for this book is available from the British Library.

Printed in Singapore

Dedicated

to Cristina, Jana and Rose

Preface

Consider the Lebesgue spaces $L_p(\Omega)$ and Sobolev spaces modelled on them: the spaces of variable integrability mentioned in the title appear in a natural way when the constant p is replaced by a function on the underlying set Ω. Our aim in this book is to give a largely self-contained account of these more general spaces. We concentrate on those aspects of the theory most useful from the standpoint of partial differential equations, while avoiding excessive generality, both in the theory and in the applications. The current high level of interest in such spaces stems largely from the fact that they provide a natural setting for the study both of partial differential equations with coefficients having variable rates of growth and the regularity of solutions of variational problems involving integrals in which the integrand satisfies non-standard growth conditions. Conspicuous among such investigations is the work of Růžička [76] involving the modelling of electrorheological fluids. As might be expected, developments involving these spaces that are driven by the needs of such concrete problems have been accompanied by a good deal of pure theory designed to explore how close is the parallel between them and the classical spaces with constant exponent p. All this has given rise to a vast literature: for detailed accounts in circumstances of considerable generality the reader may consult the books by Cruz-Uribe and Fiorenza [18] and Diening et al. [23]. Here our aim is modest: we wish to report on very recent developments concerning the p-Laplacian (p being a function) and its associated eigenvalues, with particular emphasis on the stability of eigenvalues with respect to perturbation of p, and so eschew generality in favour of those aspects of the theory most useful in this connection. As a contribution to approximation theory, we also present sharp results concerning various s-numbers, both for a Sobolev embedding and for the (weighted) Hardy operator, in each case when a variable exponent is involved.

To give a more precise impression of what is done, we need some terminology. Let Ω be a measurable subset of R^n with Lebesgue n-measure $|\Omega|$ (finite or infinite), let $\mathcal{P}(\Omega)$ be the set of all measurable functions $p : \Omega \to [1, \infty]$ and for each such p put

$$\Omega_0 = \{x \in \Omega : 1 < p(x) < \infty\}, \; \Omega_1 = \{x \in \Omega : p(x) = 1\}$$

and

$$\Omega_\infty = \{x \in \Omega : p(x) = \infty\}.$$

When $|\Omega_0| > 0$, write

$$p_- = \underset{x \in \Omega_0}{\text{ess inf}} \; p(x), \; p_+ = \underset{x \in \Omega_0}{\text{ess sup}} \; p(x);$$

otherwise set $p_- = p_+ = 1$. For every measurable scalar-valued function f on Ω we define

$$\rho_{p,\Omega}(f) = \int_{\Omega_0 \cup \Omega_1} |f(x)|^{p(x)} \, dx + \underset{x \in \Omega_\infty}{\text{ess sup}} |f(x)|$$

and put

$$\|f\|_{p,\Omega} = \inf \left\{ \lambda > 0 : \rho_p(f/\lambda) \le 1 \right\}.$$

The function $\rho_{p,\Omega}$ is a convex modular; we shall simply write ρ_p and $\|\cdot\|_p$ if there is little chance of ambiguity. It turns out that $\|\cdot\|_{p,\Omega}$ is a norm (of Luxemburg type) on the set

$$L_{p(\cdot)}(\Omega) := \left\{ f : \rho_p(f/\lambda) < \infty \text{ for some } \lambda < \infty \right\}.$$

When endowed with this norm, $L_{p(\cdot)}(\Omega)$ becomes a Banach space, often called a generalised Lebesgue space or a space with variable exponent; when p is constant it coincides (with equality of norms) with the classical Lebesgue space $L_p(\Omega)$; and even in the case of general $p \in \mathcal{P}(\Omega)$, for simplicity of notation, it will usually be denoted by $L_p(\Omega)$ if the context makes it clear what is intended. The Sobolev space $W^1_{p(\cdot)}(\Omega)$ modelled on $L_{p(\cdot)}(\Omega)$ is the family of all $f \in L_{p(\cdot)}(\Omega)$ such that all of the generalised derivatives $D_i f$ $(i = 1, ..., n)$ belong to $L_{p(\cdot)}(\Omega)$; it becomes a Banach space when furnished with the norm

$$\|f\|_{1,p(\cdot),\Omega} := \|f\|_{p(\cdot),\Omega} + \||\nabla f|\|_{p(\cdot),\Omega}.$$

As before, this Sobolev space and its norm will often be denoted simply by $W^1_p(\Omega)$ and $\|\cdot\|_{1,p,\Omega}$ respectively: the space arises naturally, for example, when considering functionals of the form $\int_\Omega |\nabla f|^{p(x)} \, dx$. Our concern in this book will be with $L_p(\Omega)$ and $W_p^1(\Omega)$: we establish some fundamental properties of these spaces, illustrate their main similarities to and differences from their classical counterparts, and then embark on a detailed study of differential operators acting on them before dealing with questions of approximation theory.

In Chapter 1, after some geometrical preliminaries, the space $L_p(\Omega)$ is introduced and some basic facts obtained. An obvious difficulty arises in connection with its norm, which is less convenient to handle than the modular $\rho_{p,\Omega}$ to which it essentially reduces when p is constant. However, all is not lost, for if $p \in \mathcal{P}(\Omega)$ is such that $1 < p_- \le p_+ < \infty$, it turns out that for all $f \in L_p(\Omega)$,

$$\min\left\{\|f\|_p^{p_-}, \|f\|_p^{p_+}\right\} \le \rho_p(f) \le \max\left\{\|f\|_p^{p_-}, \|f\|_p^{p_+}\right\}.$$

These inequalities often simplify arguments by making it possible to work with the modular instead of the norm. To begin with, results emphasising the similarities with classical Lebesgue spaces are derived: Hölder's inequality is obtained; when $p_+ < \infty$, the set of all bounded functions is proved to be dense in $L_p(\Omega)$, and when $p_+ < \infty$ the dual of $L_p(\Omega)$ is shown to be $L_{p'}(\Omega)$; the equivalence of reflexivity and uniform convexity of $L_p(\Omega)$ is established, each being equivalent to the conditions $1 < p_- \le p_+ < \infty$. Embeddings are studied when $|\Omega| < \infty$: it is shown that when $p, q \in \mathcal{P}(\Omega)$, then $L_q(\Omega)$ is continuously embedded in $L_p(\Omega)$ if and only if $p(x) \le q(x)$ a.e. in Ω. A particularly useful result is that if $p, q_k \in \mathcal{P}(\Omega)$, $\varepsilon_k > 0$ $(k \in \mathbb{N})$ are such that $1 < p_- \le p_+ < \infty$, $\lim_{k \to \infty} \varepsilon_k = 0$ and $p(x) \le q_k(x) \le p(x) + \varepsilon_k$ $(x \in \Omega, k \in \mathbb{N})$, then $\lim_{k \to \infty} \|id_k\| = 1$, where id_k is the natural embedding of $L_{q_k}(\Omega)$ in $L_p(\Omega)$. With an eye to the later chapters, the duality maps on $L_p(\Omega)$ are determined. So far, so good: the spaces with variable exponent do not seem to be exotically different from their classical counterparts. However, the first real clue that they might be rather alien in character comes when it is noticed that when p is continuous but not constant, then there is an element of $L_p(\Omega)$ that is not p-mean continuous. This lack of invariance under translation is responsible for serious problems when studying convolutions or the Hardy-Littlewood maximal operator. It was to ameliorate the position as regards such difficulties that conditions of the form

$$|p(x) - p(y)| \le -C/\log|x - y| \quad (|x - y| \text{ small and positive})$$

were introduced.

The next chapter introduces the first-order Sobolev space $W_p^1(\Omega)$ modelled on $L_p(\Omega)$. After deriving some basic properties, the focus is on inequalities of Poincaré type and analogues of the classical Sobolev embedding theorems.

Chapter 3 is concerned with the p-Laplacian Δ_p defined by

$$\Delta_p u = -\operatorname{div}\left(|\nabla u|^{p-2}\nabla u\right)$$

for all $u \in \overset{\circ}{W}{}_p^1(\Omega)$, the closure of $C_0^\infty(\Omega)$ in $W_p^1(\Omega)$. Mapping properties are established; in particular we prove the existence and uniqueness theorem of Fan and Zhang relating to the Dirichlet problem for the p-Laplacian with variable p. Then, attention is directed to the behaviour of solutions of the Dirichlet problem for Δ_p when p is varied. In the next chapter the focus turns to the eigenvalue problem for the p-Laplacian. The straighforward generalization of the eigenvalue problem for the p-Laplacian consisting of simply considering p as a function of x in its classical formulation does not give the Euler-Lagrange equation resulting from the minimization of the Rayleigh quotient. This is another manifestation of the lack of homogeneity of the modular in the variable case. In its train comes the additional problem of finding an adequate notion of an eigenvalue for the modular case. One possibility (explored in [43]) consists of employing the Ljusternik-Schirelmann theory. Here we follow a direct variational approach to define the first eigenvalue in a way that is consistent with the classical notion. We then concentrate on estimates for the change in this first eigenvalue when the integrability index p is varied. The corresponding behaviour of the eigenfunctions is also analysed.

The last chapter focuses on the study of approximation on L_p spaces and on associated Sobolev spaces. First, a review is given of the definitions and basic properties of s-numbers and n-widths. The relations between different s-numbers (including the approximation, Gelfand, Bernstein, Kolmogorov and Mityagin numbers) are described. Then the behaviour of "strict" s-numbers (that is, the approximation, Gelfand, Kolmogorov and Bernstein numbers) is studied for the Sobolev embedding E_0 with variable p on the compact interval $I = [a, b] \subset \mathbb{R}$:

$$E_0 : \overset{\circ}{W}{}_p^1(I) \to L_p(I);$$

note that functions f in the domain space have the property that $f(a) = f(b) = 0$. Under appropriate conditions on p it is shown that all the "strict" s-numbers of the compact map E_0 coincide and that in fact we have

$$\lim_{n\to\infty} n s_n(E_0) = \frac{1}{2\pi} \int_I \left(p'(x)p(x)^{p(x)-1}\right)^{1/p(x)} \sin(\pi/p(x))dx,$$

where $s_n(E_0)$ denotes any n^{th} "strict" s-number of E_0. Turning to the weighted Hardy operator $T_{v,u} : L_p(I) \to L_p(I)$, where

$$T_{v,u}f(x) := v(x) \int_a^x f(t)u(t)dt \ (x \in I)$$

and $u \in L_p(I)$, $v \in L_{p'}(I)$, techniques related to those used to deal with E_0 show that under appropriate conditions on p we have the asymptotic formula

$$\lim_{n \to \infty} ns_n(T_{v,u}) = \frac{1}{2\pi} \int_I |v(x)u(x)| \left(p'(x)p(x)^{p(x)-1} \right)^{1/p(x)} \sin(\pi/p(x))dx,$$

where $s_n(T_{v,u})$ stands for any "strict" s-number of $T_{v,u}$.

We hope that the results obtained in this book will motivate further investigation of the properties of operators on spaces with variable integrability.

It is a pleasure to express our thanks to Ying-Oi Chiew and Lai Fun Kwong for all their help in the production of this book.

Contents

Chapter 1

Preliminaries

Here we provide basic information about the geometry of Banach spaces and spaces with variable exponent. Proofs of absolutely standard results are often eschewed in favour of references to works in which full details are given.

The norm on a normed linear space X will usually be written as $\|\cdot \mid X\|$, $\|\cdot\|_X$ or even $\|\cdot\|$ if the context is clear; often $\|\cdot\|_p$ or $\|\cdot\|_{p,I}$ will be used to represent the norm on $L_p(I)$. The dimension of X is denoted by $\dim X$; and B_X (resp. S_X) will stand for the closed unit ball (resp. the unit sphere) in X; $B(x,r)$ will denote the open ball in X with centre x and radius r. By $B(X,Y)$ will be meant the space of all bounded linear maps from a Banach space X to another such space Y, written as $B(X)$ when $Y = X$ and as X^* (the dual of X) when Y is the space of scalars associated with X; the value of $x^* \in X^*$ at $x \in X$ is denoted by $\langle x, x^* \rangle_X$ or $\langle x, x^* \rangle$. Weak convergence in X is represented by a half arrow $\rightharpoonup$, weak*-convergence in X^* by $\overset{*}{\rightharpoonup}$. If X is continuously embedded in Y we write $X \hookrightarrow Y$. In the case when X is compactly embedded into Y we denote this by $X \hookrightarrow\hookrightarrow Y$. The characteristic function of a set S is written as χ_S; and the Lebesgue n-measure of a subset Ω of $\mathbb{R}^n$ will be denoted by $|\Omega|_n$, or $|\Omega|$ if there is no ambiguity.

1.1 The Geometry of Banach Spaces

Definition 1.1. A Banach space X is said to be **strictly convex** if whenever $x, y \in X$ are such that $x \neq y$ and $\|x\| = \|y\| = 1$, and $\lambda \in (0,1)$, then $\|\lambda x + (1-\lambda)y\| < 1$.

Thus no sphere in X contains a line segment. Examples of such spaces are provided by l_p and L_p, when $1 < p < \infty$; if p is 1 or ∞ this property is lost. Strict convexity is inherited by linear subspaces endowed with the induced norm. A most useful condition equivalent to strict convexity is given by the next result.

Proposition 1.1. *Let X be a Banach space. Then X is strictly convex if and only if given any $x^* \in X^* \backslash \{0\}$, there exists at most one $x \in X$ such that $\|x\| = 1$ and $\langle x, x^* \rangle = \|x^*\|$; such an x exists if X is reflexive.*

Proof.   Let X be strictly convex and $x^* \in X^* \backslash \{0\}$. Suppose there are two distinct such points x, say x_1 and x_2. Then if $0 < \lambda < 1$,

$$\|x^*\| = \lambda \langle x_1, x^* \rangle + (1 - \lambda) \langle x_2, x^* \rangle = \langle \lambda x_1 + (1 - \lambda) x_2, x^* \rangle$$
$$\leq \|x^*\| \, \|\lambda x_1 + (1 - \lambda) x_2\| < \|x^*\|,$$

which is absurd. For the converse, suppose that $\|x + \lambda(y - x)\| = 1$ for some $x, y \in S_X$ and some $\lambda \in (0, 1)$. By the Hahn-Banach theorem, there exists $x^* \in X^*$ such that $\langle x + \lambda(y - x), x^* \rangle = 1$ and $\|x^*\| = 1$. Then $(1 - \lambda) \langle x, x^* \rangle + \lambda \langle y, x^* \rangle = 1$, and since $|\langle x, x^* \rangle|, |\langle y, x^* \rangle| \leq 1$ we must have $\langle x, x^* \rangle = \langle y, x^* \rangle = 1$. By hypothesis, this implies that $x = y$, and so X is strictly convex.

To prove that such an x exists when X is reflexive, let (x_k) be a sequence in S_X such that $\|x^*\| = \lim_{k \to \infty} \langle x_k, x^* \rangle$. Then there is a weakly convergent subsequence of (x_k), still denoted by (x_k) for simplicity, with weak limit x, say. Hence $\|x\| \leq 1$ and $\langle x, x^* \rangle = \lim_{k \to \infty} \langle x_k, x^* \rangle = \|x^*\|$.   □

The next result indicates how strict convexity can be used.

Proposition 1.2. *Let K be a closed, convex, non-empty subset of a strictly convex Banach space X. Then there is at most one element $x \in K$ such that*

$$\|x\| = \inf\{\|y\| : y \in K\}.$$

If X is reflexive, such an x exists.

Proof.   Suppose there exist $x, y \in K$ with $\|x\| = \|y\| = \inf\{\|z\| : z \in K\}$, $x \neq y$. Let $0 < \lambda < 1$: then $\lambda x + (1 - \lambda)y \in K$, $\|\lambda x + (1 - \lambda)y\| < \|x\|$ and we have a contradiction.

Now let X be reflexive and assume that (x_k) is a sequence in K such that $\lim_{k \to \infty} \|x_k\| = l := \inf\{\|y\| : y \in K\}$. This sequence has a subsequence, again denoted by (x_k), such that $x_k \rightharpoonup x$ for some $x \in X$; and $x \in K$ since K is convex and closed, and hence weakly closed. Since $\|x\| \leq \lim_{k \to \infty} \|x_k\| = l$, the result follows.   □

We now introduce an important class of strictly convex spaces.

Definition 1.2. The **modulus of convexity** $\delta_X : (0,2] \to [0,1]$ of a Banach space X with $\dim X \geq 2$ is defined by

$$\delta_X(\varepsilon) = \inf \left\{ 1 - \frac{1}{2} \|x + y\| : x, y \in B_X, \|x - y\| \geq \varepsilon \right\}.$$

The space X is called **uniformly convex** if $\delta_X(\varepsilon) > 0$ for all $\varepsilon \in (0,2]$.

Remark 1.1. Some useful basic facts concerning uniform convexity are listed below: for details of the proofs of these assertions see [38], Chapter 9 and [60] II, Chapter 1.

(i) The modulus of convexity δ_X is increasing on $[0,2]$, continuous on $[0,2)$ (not necessarily at 2) and $\delta_X(0) = 0$; also $\delta_X(2) = 1$ if and only if X is strictly convex. Moreover,

$$\delta_X(\varepsilon) = \inf \left\{ \delta_Y(\varepsilon) : Y \text{ is a 2-dimensional subspace of } X \right\}.$$

(ii) An equivalent formula for the modulus of convexity of X is

$$\delta_X(\varepsilon) = \inf \left\{ 1 - \frac{1}{2} \|x + y\| : x, y \in S_X, \|x - y\| = \varepsilon \right\}.$$

(iii) Each of the following two conditions is equivalent to the hypothesis of uniform convexity of X :
  (a) If (x_n), (y_n) are sequences in X such that (x_n) is bounded and

$$\lim_{n \to \infty} \left(2 \|x_n\|^2 + 2 \|y_n\|^2 - \|x_n + y_n\|^2 \right) = 0,$$

then $\lim_{n \to \infty} \|x_n - y_n\| = 0$.
  (b) If for all $n \in \mathbb{N}$ there are points $x_n, y_n \in B_X$ such that $\lim_{n \to \infty} \|x_n + y_n\| = 2$, then $\lim_{n \to \infty} \|x_n - y_n\| = 0$.

(iv) Every closed linear subspace of a uniformly convex space is uniformly convex when given the inherited norm. While every uniformly convex space is strictly convex, the converse is false, in general. However, if X is strictly convex and $\dim X < \infty$, then X is uniformly convex.

(v) Every Hilbert space H is uniformly convex, and

$$\delta_H(\varepsilon) \geq \varepsilon^2/8.$$

For every Banach space X of dimension at least 2,

$$\delta_X(\varepsilon) \leq \delta_H(\varepsilon) = 1 - (1 - \varepsilon^2/4)^{1/2} \leq C\varepsilon^2.$$

In this sense, Hilbert spaces are the 'most' uniformly convex spaces.

(vi) Consider the Lebesgue space $L_p = L_p(\Omega, \mu)$, where $p \in (1, \infty)$, (Ω, μ) is a measure space and the norm is given by

$$\|x\|_p := \left(\int_\Omega |x|^p \, d\mu \right)^{1/p}.$$

The modulus of convexity of L_p is denoted by δ_p. Each such space L_p is uniformly convex, and for all $\varepsilon \in (0, 2)$,

$$\delta_p(\varepsilon) \geq \begin{cases} 1 - \left\{ 1 - (p-1)(\varepsilon/2)^2 \right\}^{1/2} \geq (p-1)\varepsilon^2/8, & 1 < p \leq 2, \\ (\varepsilon/2)^p/p, & 2 \leq p < \infty. \end{cases}$$

Thus if $1 < p < \infty$, then the sequence space l_p and its n-dimensional analogue l_p^n $(n \in \mathbb{N})$ are uniformly convex. Since $\prod_{k=1}^n L_p$, endowed with the norm

$$\|(f_1, ..., f_n)\| := \left(\sum_{k=1}^n \|f_k\|_p^p \right)^{1/p},$$

may be thought of as a space of l_p^n form, it can be seen that $\prod_{k=1}^n L_p$ is also uniformly convex, with the same modulus of convexity as L_p.

(vii) Every uniformly convex Banach space is reflexive.

(viii) If (x_k) is a sequence in a uniformly convex space X that converges weakly to $x \in X$, with $\|x_k\| \to \|x\|$, then $\|x_k - x\| \to 0$.

Next we consider projections. Let K be a closed, convex, non-empty subset of a uniformly convex Banach space X. Given any $x \in X$, the set $K - \{x\} := \{y - x : y \in K\}$ is also closed, convex and non-empty, and so by Proposition 1.2, it has a unique element $w(x)$ of minimal norm. Since $w(x) = z(x) - x$ for some unique $z(x) \in K$, we have

$$\|z(x) - x\| = \inf\{\|y - x\| : y \in K\} = d(x, K).$$

Hence $z(x)$ is the unique point of K nearest to x : we write $P_K x = z(x)$ and call the map $P_K : X \to K$ thus defined the **projection of X onto K**. Clearly $P_K x = x$ if and only if $x \in K$. This map is continuous and, in general, is nonlinear. If X is a Hilbert space, it is easy to show that P_K is Lipschitz-continuous:

$$\|P_K x - P_K y\| \leq \|x - y\| \text{ for all } x, y \in X.$$

Proofs of these assertions are given in [31], 1.2.

A characterisation of strict and uniform convexity can be given by means of differentiability considerations. Let U be an open subset of a Banach space X and $f : U \to \mathbb{R}$. We say that f is **Gâteaux-differentiable** at $x_0 \in U$ if there exists $x^* \in X^*$ such that

$$\langle h, x^* \rangle = \lim_{t \to 0} \frac{f(x_0 + th) - f(x_0)}{t} \text{ for all } h \in X.$$

The limit above is called the **derivative of f in the direction h**; the functional x^* is denoted by grad $f(x_0)$ and will be referred to as the **gradient** or **Gâteaux derivative** of f at x_0. If this limit is uniform with respect to $h \in S_X$, we say that f is **Fréchet-differentiable** at x_0 (and then refer to x^* as the **Fréchet derivative** of f at x_0, written $f'(x_0)$): equivalently,

$$\frac{f(x_0 + h) - f(x_0) - \langle h, x^* \rangle}{\|h\|} \to 0 \text{ as } \|h\| \to 0.$$

Clearly Fréchet-differentiability implies Gâteaux-differentiability; in the reverse direction, it can be shown that if grad $f(x)$ exists throughout some neighbourhood of x_0 and is continuous at x_0, then f is Fréchet-differentiable at x_0 and grad $f(x_0) = f'(x_0)$.

The case when $f(x) = \|x\|$ $(x \in X)$ is of particular interest. The norm $\|\cdot\|$ is said to be Fréchet- (resp. Gâteaux-) differentiable if $\|\cdot\|$ is Fréchet- (resp. Gâteaux-) differentiable at every point of $X \backslash \{0\}$: the point 0 is excluded because no norm is differentiable at 0. Given a convex subset C of a Banach space X and a point $x \in C$, a functional $f \in X^*$ is called a **supporting functional of C at x** if $\|f\| = 1$ and $\langle x, f \rangle = \sup \{ |\langle y, f \rangle| : y \in C \}$. It emerges that if the norm on X is Gâteaux-differentiable, then grad $\|x\|$ is a supporting functional of B_X at every point x of $B_X \backslash \{0\}$: for a proof of this and of the characterisations given next of spaces with Gâteaux-differentiable norms we refer to [38], Chapter 8.

Theorem 1.1. *The following statements about a Banach space X are equivalent:*
(i) The norm on X is Gâteaux-differentiable.
(ii) The dual space X^* is strictly convex.
(iii) Given any $x \in S_X$, there is a unique $f \in S_{X^*}$ such that $\langle x, f \rangle = 1$.

It turns out that uniform convexity of a space X has significant implications for the dual space X^*. To explain this we introduce a new object, the modulus of smoothness of a space.

Definition 1.3. Let X be a Banach space. The **modulus of smoothness** of X is the function $\rho_X : (0, \infty) \to [0, \infty)$ defined by

$$\rho_X(\tau) = \sup \left\{ \frac{\|x + \tau h\| + \|x - \tau h\|}{2} - 1 : x, h \in S_X \right\}.$$

If $\lim_{\tau \to 0} \rho_X(\tau)/\tau = 0$, the space X is said to be **uniformly smooth**.

Note that ρ_X really is a non-negative function since for all $x, h \in X$ we have $2\|x\| = \|x + \tau h + x - \tau h\| \leq \|x + \tau h\| + \|x - \tau h\|$. Evidently the property of uniform smoothness is preserved on passage to a subspace. For the following theorem see, for example, [38], Chapter 8.

Lemma 1.1. *Let X be a Banach space. The following statements are equivalent:*

(i) The space X is uniformly smooth.

(ii) The limit

$$\lim_{t \to 0} \frac{\|x + \tau h\| - \|x\|}{t} = \langle h, \mathrm{grad}\, \|x\| \rangle$$

exists, uniformly for $x, h \in S_X$.

(iii) The norm of X is Fréchet-differentiable on S_X and the map $x \longmapsto \mathrm{grad}\, \|x\| : S_X \to S_{X^*}$ is uniformly continuous.

If $\|\cdot\|$ satisfies any of the equivalent conditions in the last lemma we shall say that it is **uniformly Fréchet-differentiable.**

Connections between the moduli of convexity and smoothness, together with important implications of these connections, are given in the next theorem.

Theorem 1.2. *Let X be a Banach space. Then:*

(i) for all $\tau > 0$,

$$\rho_{X^*}(\tau) = \sup \left\{ \frac{1}{2}\tau\varepsilon - \delta_X(\varepsilon) : 0 < \varepsilon \leq 2 \right\}$$

and

$$\rho_X(\tau) = \sup \left\{ \frac{1}{2}\tau\varepsilon - \delta_{X^*}(\varepsilon) : 0 < \varepsilon \leq 2 \right\};$$

(ii) X is uniformly convex if and only if X^* is uniformly smooth;

(iii) X is uniformly smooth if and only if X^* is uniformly convex.

Proof. (i) We establish only the first of the two statements, the proof of the second being similar. Let $\varepsilon \in (0, 2]$ and $\tau > 0$; let $x, y \in S_X$ be such that $\|x - y\| \geq \varepsilon$. By the Hahn-Banach theorem, there exist $f, g \in S_{X^*}$ such that $\langle x + y, f \rangle = \|x + y\|$ and $\langle x - y, g \rangle = \|x - y\|$. Then

$$2\rho_{X^*}(\tau) \geq \|f + \tau g\| + \|f - \tau g\| - 2$$
$$\geq \mathrm{re}\, \langle x, f + \tau g \rangle + \mathrm{re}\, \langle y, f - \tau g \rangle - 2$$
$$= \langle x + y, f \rangle + \tau \langle x - y, g \rangle - 2 = \|x + y\| + \tau \|x - y\| - 2.$$

Thus $2 - \|x + y\| \geq \tau\varepsilon - 2\delta_{X^*}(\tau)$, which gives $\delta_X(\varepsilon) + \rho_{X^*}(\tau) \geq \tau\varepsilon/2$, so that $\rho_{X^*}(\tau) \geq \sup\left\{\frac{1}{2}\tau\varepsilon - \delta_X(\varepsilon) : 0 < \varepsilon \leq 2\right\}$.

To establish the reverse inequality, let $\tau > 0$ and $f, g \in S_{X^*}$. Given $\eta > 0$, let $x, y \in S_X$ be such that re $\langle x, f + \tau g \rangle \geq \|f + \tau g\| - \eta$ and re $\langle x, f - \tau g \rangle \geq \|f - \tau g\| - \eta$. Then

$$
\begin{aligned}
\|f + \tau g\| + \|f - \tau g\| - 2 &\leq \text{ re } \langle x + y, f \rangle + \tau \text{ re } \langle x - y, g \rangle - 2 + 2\eta \\
&\leq \|x + y\| - 2 + \tau \|x - y\| + 2\eta \\
&\leq -2\delta_X(\|x - y\|) + \tau \|x - y\| + 2\eta \\
&\leq 2\sup\left\{\frac{1}{2}\tau\varepsilon - \delta_{X^*}(\varepsilon) : 0 < \varepsilon \leq 2\right\} + 2\eta,
\end{aligned}
$$

from which the desired inequality follows.

(ii) Suppose that X is uniformly convex and let $\varepsilon_0 \in (0, 2]$. Then $\delta_X(\varepsilon) \geq \delta_X(\varepsilon_0) > 0$ if $\varepsilon \in [\varepsilon_0, 2]$. Let $\tau \in (0, \delta_X(\varepsilon_0))$. If $\varepsilon \in [\varepsilon_0, 2]$, then

$$
\varepsilon/2 - \delta_X(\varepsilon)/\tau \leq \varepsilon/2 - \delta_X(\varepsilon_0)/\tau \leq \varepsilon/2 - 1 \leq 0.
$$

By (i),

$$
\rho_{X^*}(\tau)/\tau = \sup_{0 < \varepsilon \leq \varepsilon_0} (\varepsilon/2 - \delta_X(\varepsilon)/\tau) \leq \sup_{0 < \varepsilon \leq \varepsilon_0} \varepsilon/2 = \varepsilon_0/2.
$$

Thus $\lim_{\tau \to 0} \rho_{X^*}(\tau)/\tau = 0 : X^*$ is uniformly smooth.

Conversely, if X is not uniformly convex, then $\delta_X(\varepsilon_0) = 0$ for some $\varepsilon_0 \in (0, 2]$. By (i), if $\tau > 0$ then

$$
\rho_{X^*}(\tau) = \sup_{0 < \varepsilon \leq \varepsilon_0} (\varepsilon\tau/2 - \delta_X(\varepsilon)) \geq \varepsilon_0\tau/2.
$$

Hence $\lim \sup_{\tau \to 0} \rho_{X^*}(\tau)/\tau \geq \varepsilon_0/2$, which implies that X^* is not uniformly smooth.

(iii) The proof is similar to that of (ii), again using (i). $\qquad\square$

Note that from the last theorem it follows that if X is uniformly smooth, it is reflexive.

For a Hilbert space H it is simple to verify that

$$
\rho_H(\tau) = \left(1 + \tau^2\right)^{1/2} - 1,
$$

so that H is uniformly smooth: of course this also follows from the fact that H is uniformly convex. It is known (see [60] II, Chapter 1e) that for every Banach space X we have

$$
\rho_X(\tau) \geq \left(1 - \tau^2\right)^{1/2} - 1.
$$

This supports the view that Hilbert spaces are the 'most uniformly smooth' as well as the 'most uniformly smooth' spaces.

To complete this section we deal with duality maps. A map μ : $[0, \infty) \rightarrow [0, \infty)$ that is continuous, strictly increasing and satisfies $\mu(0) = 0$, $\lim_{t \to \infty} \mu(t) = \infty$, is called a **gauge function**. A map J from a Banach space X to 2^{X^*}, the set of all subsets of X^*, is said to be a **duality map on X with gauge function** μ if for all $x \in X$,

$$J(x) = \{x^* \in X^* : \langle x, x^* \rangle = \|x^*\| \, \|x\|, \|x^*\| = \mu(\|x\|)\}.$$

By the Hahn-Banach theorem, for each $x \in X$ the set $J(x)$ is non-empty; it is also convex. For if $x^*, y^* \in J(x)$ and $\lambda \in (0, 1)$, $z^* = \lambda x^* + (1 - \lambda) y^*$, then

$$\langle x, z^* \rangle = \lambda \langle x, x^* \rangle + (1 - \lambda) \langle x, y^* \rangle = \mu(\|x\|) \|x\|.$$

Hence $\|z^*\| \geq \mu(\|x\|)$. Since

$$\|z^*\| \leq \lambda \|x^*\| + (1 - \lambda) \|y^*\| = \mu(\|x\|),$$

we see that $\|z^*\| = \mu(\|x\|)$, which means that $z^* \in J(x)$ and establishes the asserted convexity.

Let X be a Banach space with strictly convex dual X^* and let J be a duality map on X with gauge function μ. Then for each $x \in X$, the set $J(x)$ consists of precisely one point. In fact, for each $x \in X$, the points in $J(x)$ lie on the sphere in X^* with centre 0 and radius $\mu(\|x\|)$. If $J(x)$ contained two distinct points, the midpoint of the line segment joining them would be in the convex set $J(x)$, which is impossible as X^* is strictly convex.

In view of this result, we shall regard a duality map J on X as a map from X to X^* when X^* is strictly convex. Duality maps have several useful properties, summarised in the next theorem (see [63], p.176), to prepare for which we introduce some terminology. Let X be a real Banach space. A map $T : X \rightarrow X^*$ is called **monotone** if

$$\langle x - y, Tx - Ty \rangle \geq 0 \text{ for all } x, y \in X;$$

if

$$\langle x - y, Tx - Ty \rangle > 0 \text{ for all } x, y \in X \text{ with } x \neq y$$

it is said to be **strictly monotone**; it is **coercive** if

$$\langle x, Tx \rangle / \|x\| \rightarrow \infty \text{ as } \|x\| \rightarrow \infty;$$

and finally it is **demicontinuous** if $x_n \rightarrow x$ implies that $Tx_n \xrightarrow{*} Tx$.

Theorem 1.3. *Let X be a real, reflexive Banach space with strictly convex dual X^* and let J be a duality map on X with gauge function μ. Then J is monotone, coercive and demicontinuous (continuous if X^* is uniformly convex); if, in addition, X is strictly convex, then J is strictly monotone.*

Proof. That J is monotone is clear since for all $x, y \in X$,

$$\langle x - y, Jx - Jy \rangle = \mu\left(\|x\|\right)\|x\| + \mu\left(\|y\|\right)\|y\| - \langle x, Jy \rangle - \langle y, Jx \rangle$$
$$\geq \left(\mu\left(\|x\|\right) - \mu\left(\|y\|\right)\right)\left(\|x\| - \|y\|\right) \geq 0.$$

As $\langle x, Jx \rangle = \mu\left(\|x\|\right)\|x\|$ $(x \in X)$ the coercivity of J is immediate. For demicontinuity, it is enough to deal with the case in which $\mu(t) = t$ $(t \geq 0)$. Let (x_n) be a sequence in X that converges to x and let $u \in X$: we must show that $\langle u, Jx_n \rangle \to \langle u, Jx \rangle$. Since the sequence (x_n) and the points x, u span a separable subspace of X it is sufficient to deal with the case in which X is separable. As $\left(\|Jx_n\|\right)$ is bounded, it follows that, by passage to a subsequence if necessary, $Jx_n \xrightarrow{*} x^*$ for some $x^* \in X^*$ (see, for example, [78], Theorem 4.41A). Thus

$$\langle x_n, Jx_n \rangle \to \langle x, x^* \rangle, \quad \|x^*\| \leq \liminf_{n \to \infty} \|Jx_n\| = \lim_{n \to \infty} \|x_n\| = \|x\|,$$

while $\langle x_n, Jx_n \rangle = \|x_n\|^2 \to \|x\|^2$. The definition of J now ensures that $x^* = Jx$. Since this argument holds for every weak*-convergent subsequence of (Jx_n), the demicontinuity of J follows. If X^* is uniformly convex, then since $Jx_n \rightharpoonup Jx$ (the reflexivity of X follows from that of X^*) and $\|Jx_n\| = \|x_n\| \to \|x\| = \|Jx\|$, it follows that $Jx_n \to Jx$.

Finally, suppose that X is strictly convex. To establish the strict monotonicity of J we must prove that $\langle x - y, Jx - Jy \rangle = 0$ implies that $x = y$. From the argument given above to show that J is monotone we see that $\|x\| = \|y\|$: suppose that $x \neq y$. Then $x/\|x\| \neq y/\|y\|$ and so by Proposition 1.2,

$$\|Jx\| = \langle x/\|x\|, Jx \rangle > \langle y/\|y\|, Jx \rangle,$$

whence $\langle y, Jx \rangle < \langle x, Jx \rangle$; similarly, $\langle x, Jy \rangle < \langle y, Jy \rangle$. Hence

$$0 = \langle x - y, Jx - Jy \rangle > \langle x, Jx \rangle + \langle y, Jy \rangle - \langle x, Jx \rangle - \langle y, Jy \rangle = 0.$$

This contradiction completes the proof. $\qquad\square$

It is natural to ask whether J is compact, that is, does J map bounded sets into relatively compact ones? If dim $X = \infty$, the answer is 'No'. To explain this we invoke the Bishop-Phelps theorem in the form that asserts that the set of elements of X^* that attain their maxima on S_X is dense in X^* (see, for example, [72], 3.4.3.9). It is sufficient to deal with the case in which the gauge function μ is normalised by the condition $\mu(1) = 1$. Let $x^* \in S_{X^*}$. By the Bishop-Phelps theorem, there are sequences $(x_n^*) \subset X^*$ and $(x_n) \subset S_X$ such that

$$x_n^* \to x^* \text{ and } \langle x_n, x_n^* \rangle = \|x_n^*\| \to \|x^*\| = 1.$$

Thus $\left\langle x_n, \frac{x_n^*}{\|x_n^*\|} \right\rangle = 1$, and so $\frac{x_n^*}{\|x_n^*\|} = Jx_n$: hence $Jx_n \to x^*$. This shows that $S_{X^*} \subset \overline{JS_X}$, so that if J were compact, S_{X^*} would be a compact subset of X^*, which is possible if and only if X^* (and hence also X) is finite-dimensional.

It is well known (see, for example, [21], Theorem 12.1) that if X is a real, reflexive Banach space and $T : X \to X^*$ is monotone, coercive and demicontinuous, then T is surjective. Together with Theorem 1.3 this gives

Corollary 1.1. *Let X be a real, reflexive Banach space that is strictly convex and has strictly convex dual; let J be a duality map on X with gauge function μ. Then J maps X surjectively onto X^*, and the map $f \longmapsto J^{-1}(f)$ is a duality map on X^* with gauge function μ^{-1} (X^{**} being identified with X).*

When X is a Hilbert space, so that X^* may be identified with X, the most natural duality map on X is the identity map, corresponding to the gauge function μ with $\mu(t) = t$. If $1 < p < \infty$ and $\mu(t) = t^{p-1}$ ($t \geq 0$), it is easy to check that the duality map J on $L_p(\Omega)$ (where Ω is, for example, a measurable subset of $\mathbb{R}^n$) is given by $J(u) = |u|^{p-2} u$; the duality map on l_p with the same gauge function is defined by $J((x_k)) = \left(|x_k|^{p-2} x_k \right)$.

1.2   Spaces with Variable Exponent

Let X be a linear space over $\mathbb{K}$ ($\mathbb{R}$ or $\mathbb{C}$). A map $\rho : X \to [0, \infty]$ is called a **convex modular** on X if
(i) $\rho(x) = 0$ if and only if $x = 0$;
(ii) $\rho(\lambda x) = \rho(x)$ for all $x \in X$ and all $\lambda \in \mathbb{K}$ with $|\lambda| = 1$;
(iii) ρ is convex: $\rho(\alpha x + \beta y) \leq \alpha \rho(x) + \beta \rho(y)$ for all $x, y \in X$ and all $\alpha, \beta \geq 0$ with $\alpha + \beta = 1$.

Note that (i) and (iii) imply that for each $x \in X$, the map $\lambda \longmapsto \rho(\lambda x)$ is non-decreasing on $[0, \infty)$; also,

$$\rho(\lambda x) = \rho(|\lambda| x) \leq |\lambda| \rho(x) \text{ if } |\lambda| \leq 1,$$
$$\rho(\lambda x) = \rho(|\lambda| x) \geq |\lambda| \rho(x) \text{ if } |\lambda| \geq 1. \tag{1.1}$$

A convex modular ρ is said to be (left-, right-) continuous if, for every $x \in X$, the map $\lambda \longmapsto \rho(\lambda x)$ is (left-, right-) continuous on $[0, \infty)$ (on $(0, \infty)$ for left-continuity).

Given any normed linear space $(X, \|\cdot\|)$, it is clear that $\|\cdot\|$ is a continuous convex modular on X. In the reverse direction, let X be a linear space and consider the subset X_ρ defined by

$$X_\rho = \{x \in X : \rho(\lambda x) < \infty \text{ for some } \lambda > 0 \},$$

where ρ is a convex modular on X. This subset is called a **modular space**; in view of (1.1),

$$X_\rho = \left\{ x \in X : \lim_{\lambda \to 0} \rho(\lambda x) = 0 \right\}.$$

Proposition 1.3. *Let ρ be a convex modular on a linear space X and define*

$$\|x\|_\rho := \inf \{\lambda > 0 : \rho(x/\lambda) \le 1\} \quad (x \in X_\rho).$$

Then:
(i) $\left(X_\rho, \|\cdot\|_\rho\right)$ *is a normed linear space;*
(ii) if $\rho(x) \le 1$, then $\|x\|_\rho \le 1$;
(iii) if ρ is left-continuous, then $\|x\|_\rho \le 1$ if and only if $\rho(x) \le 1$. If ρ is continuous, then $\|x\|_\rho < 1$ if and only if $\rho(x) < 1$; and $\|x\|_\rho = 1$ if and only if $\rho(x) = 1$.

Proof. (i) Let $x, y \in X_\rho$ and $\lambda \in \mathbb{K}$. From (1.1) it follows that $\lambda x \in X_\rho$. Since ρ is convex,

$$\rho(\lambda(x+y)) \le \frac{1}{2} \{\rho(2\lambda x) + \rho(2\lambda y)\} \to 0 \text{ as } \lambda \to 0,$$

and so $x + y \in X_\rho$. Thus X_ρ is a linear space.

To show that $\|\cdot\|_\rho$ is a norm, note first that plainly $\|x\|_\rho < \infty$ for all $x \in X_\rho$, and $\|0\|_\rho = 0$. Let $x, y \in X_\rho$ and $\lambda \in \mathbb{K}$. Then

$$\|\lambda x\|_\rho = \inf \{\beta > 0 : \rho(\lambda x/\beta) \le 1\} = |\lambda| \inf \{\beta > 0 : \rho(x/\beta) \le 1\} = |\lambda| \|x\|_\rho.$$

Now let $\lambda > \|x\|_\rho$ and $\mu > \|y\|_\rho$, so that $\rho(x/\lambda) \le 1$ and $\rho(y/\mu) \le 1$. By the convexity of ρ,

$$\rho\left(\frac{x+y}{\lambda+\mu}\right) \le \frac{\lambda}{\lambda+\mu} \rho\left(\frac{x}{\lambda}\right) + \frac{\mu}{\lambda+\mu} \rho\left(\frac{y}{\mu}\right) \le 1.$$

Thus $\|x + y\|_\rho \le \lambda + \mu$, and the triangle inequality follows. Finally, if $\|x\|_\rho = 0$, then $\rho(\alpha x) \le 1$ for all $\alpha > 0$, and hence, for all $\lambda > 0$ and $\beta \in (0, 1]$,

$$\rho(\lambda x) \le \beta \rho(\lambda x/\beta) \le \beta.$$

This shows that $\rho(\lambda x) = 0$ for all $\lambda > 0$, and so $x = 0$.
(ii) If $\rho(x) \le 1$, then from the definition of $\|x\|_\rho$ it is immediate that $\|x\|_\rho \le 1$.

(iii) Suppose that ρ is left-continuous and that $\|x\|_\rho \leq 1$. Then for all $\mu \in (0,1)$ we have $\rho(\mu x) \leq 1$. Then $\rho(x) = \lim_{\mu \to 1-} \rho(\mu x) \leq 1$.

Now assume that ρ is continuous. If $\|x\|_\rho < 1$, then for some $\lambda < 1$ we have $\rho(x/\lambda) < 1$, so that in view of the convexity of ρ,

$$\rho(x) \leq \lambda \rho(x/\lambda) \leq \lambda < 1.$$

Conversely, if $\rho(x) < 1$, then by the continuity of ρ, $\rho(\gamma x) < 1$ for some $\gamma > 1$. Thus $\|\gamma x\|_\rho \leq 1$ and $\|x\|_\rho \leq 1/\gamma < 1$. The rest is obvious. $\square$

The norm $\|\cdot\|_\rho$ is called the **Luxemburg norm** on X_ρ.

Corollary 1.2. *Let ρ be a left-continuous modular on a linear space X. Then*
(i) if $\|x\|_\rho \leq 1$, then $\rho(x) \leq \|x\|_\rho$; if $\|x\|_\rho > 1$, then $\rho(x) \geq \|x\|_\rho$;
(ii) for all $x \in X$, $\|x\|_\rho \leq \rho(x) + 1$.

Proof.   (i) Suppose $0 < \|x\|_\rho \leq 1$. Since $\left\|x/\|x\|_\rho\right\|_\rho = 1$, it follows from Proposition 1.3 that $\rho\left(x/\|x\|_\rho\right) \leq 1$; as ρ is convex, $\rho(x)/\|x\|_\rho \leq 1$. On the other hand, if $\|x\|_\rho > 1$, then $\rho(x/\lambda) > 1$ if $1 < \lambda < \|x\|_\rho$. Convexity now gives $1 < \rho(x)/\lambda$, so that $\rho(x) \geq \|x\|_\rho$.
(ii)   This is clear from (i). $\square$

Any convex, continuous, strictly increasing function $\phi : [0,\infty) \to [0,\infty)$ such that $\phi(0) = 0$ and $\lim_{t\to\infty} \phi(t) = \infty$ is called an **Orlicz function**. Examples of such functions are provided by the maps with values t^p $(1 \leq p < \infty)$, $t\log(1+t)$, $\exp(t) - 1$ at $t \in [0,\infty)$. Let Ω be a measurable subset of $\mathbb{R}^n$ and let $\mathcal{M}(\Omega)$ be the set of all Lebesgue-measurable, scalar-valued functions on Ω, functions equal almost everywhere being identified. If ϕ is an Orlicz function, the function ρ defined by

$$\rho(f) = \int_\Omega \phi(|f(x)|)\,dx \quad (f \in \mathcal{M}(\Omega))$$

is plainly a convex modular on $\mathcal{M}(\Omega)$. In fact, it is left-continuous, for if $\lambda \uparrow 1$, then $\phi(\lambda|f(x)|) \uparrow \phi(|f(x)|)$, and so by monotone convergence, $\lim_{\lambda \to 1-} \rho(\lambda f) = \rho(f)$. The corresponding modular space, denoted by $L_\phi(\Omega)$, is called an Orlicz space:

$$L_\phi(\Omega) = \{f \in \mathcal{M}(\Omega) : \rho(\lambda f) < \infty \text{ for some } \lambda > 0\};$$

the Luxemburg norm on $L_\phi(\Omega)$ is denoted by $\|\cdot \mid L_\phi(\Omega)\|$ and is given by

$$\|f \mid L_\phi(\Omega)\| = \inf\left\{\lambda > 0 : \int_\Omega \phi(|f(x)|)\,dx \leq 1\right\}.$$

A norm on $L_\phi(\Omega)$ equivalent to $\|\cdot \mid L_\phi(\Omega)\|$ is given by means of the so-called **complementary function** (to ϕ). This is the function $\phi_* : [0, \infty) \to [0, \infty)$ defined by

$$\phi_*(s) = \sup \{st - \phi(t) : t \geq 0\};$$

it is assumed that ϕ has a right-derivative that is zero at 0 and tends to infinity at infinity. Then ϕ_* is a non-degenerate Orlicz function, corresponding to which the Orlicz norm $\|\cdot\|^0$ on $L_\phi(\Omega)$ is defined by

$$\|f \mid L_\phi(\Omega)\|^0 = \sup \left\{ \left| \int_\Omega f(t)g(t)dt \right| : g \in L_{\phi_*}(\Omega), \int_\Omega \phi_*\left(|g(x)|\right) dx \leq 1 \right\};$$

moreover,

$$\|f \mid L_\phi(\Omega)\| \leq \|f \mid L_\phi(\Omega)\|^0 \leq 2 \|f \mid L_\phi(\Omega)\|$$

for all $f \in L_\phi(\Omega)$. Note that ϕ is complementary to ϕ_* : we shall therefore refer to ϕ, ϕ_* as a complementary pair.

Hölder's inequality holds in the form

$$\int_\Omega |f(t)g(t)| \, dt \leq \|f \mid L_\phi(\Omega)\|^0 \|g \mid L_{\phi^*}(\Omega)\|$$

for all $f \in L_\phi(\Omega)$ and $g \in L_{\phi_*}(\Omega)$. For proofs of these assertions see [11], 4.8, [56], Theorem 3.7.5 and [60] I,4.

The Orlicz spaces just introduced are particular cases of Orlicz-Musielak spaces, which we now define. Let $M : \Omega \times [0, \infty) \to [0, \infty)$ be such that $M(\cdot, t)$ is measurable for every $t \geq 0$, and for almost all $x \in \Omega$, the function $M(x, \cdot)$ is an Orlicz function. The Orlicz-Musielak space $L_M(\Omega)$ is defined by

$$L_M(\Omega) = \left\{ f \in \mathcal{M}(\Omega) : \int_\Omega M\left(x, \lambda \, |f(x)|\right) dx < \infty \text{ for some } \lambda > 0 \right\};$$

endowed with the norm

$$\|f \mid L_M(\Omega)\| := \inf \left\{ \lambda > 0 : \int_\Omega M\left(x, |f(x)| / \lambda\right) dx \leq 1 \right\},$$

it is a Banach space (see [70]).

It is now convenient to introduce the notion of a Banach function space. A linear space $X \in \mathcal{M}(\Omega)$ is called a Banach function space if there is a map $\|\cdot\|_X : \mathcal{M}(\Omega) \to [0, \infty]$ with the properties of a norm and such that

(i) $f \in X$ if and only if $\|f\|_X < \infty$;

(ii) $\|f\|_X = \|\,|f|\,\|_X$ for all $f \in \mathcal{M}(\Omega)$;

(iii) if $0 \leq f_k \uparrow f$, then $\|f_k\|_X \uparrow \|f\|_X$;

(iv) if $E \subset \Omega$ and $|E| < \infty$, then $\|\chi_E\|_X < \infty$;

(v) if $E \subset \Omega$ and $|E| < \infty$, there is a constant $c(E)$ such that for all $f \in X$,

$$\int_E |f(x)|\,dx \leq c(E)\,\|f\|_X\,.$$

The distribution function of an element f of a Banach function space X is the map $\mu_f : [0, \infty) \to [0, \infty]$ defined, for all $\lambda \geq 0$, by

$$\mu_f(\lambda) = |\{x \in \Omega : |f(x)| > \lambda\}|_n\,.$$

Two functions $f, g \in X$ are said to be equimeasurable if their distribution functions coincide; X is called rearrangement-invariant if $\|f\|_X = \|g\|_X$ whenever f and g are equimeasurable.

Every Banach function space X is a Banach space when endowed with the norm $\|\cdot\|_X$ (see [11], p. 6). It is routine to check that the classical Lebesgue and Sobolev spaces are Banach function spaces. While every Orlicz space $L_\phi(\Omega)$ is rearrangement-invariant (see [2], p. 270 and [10], p. 75), this is not the case, in general, for the Orlicz-Musielak spaces $L_M(\Omega)$.

Given a Banach function space X, the set

$$X' := \left\{ f \in \mathcal{M}(\Omega) : \int_\Omega |f(x)g(x)|\,dx < \infty \text{ for all } g \in X \right\},$$

furnished with the norm

$$\|f\|_{X'} := \sup_{g \in S_X} \int_\Omega |f(x)g(x)|\,dx,$$

is a Banach function space called the **associate space** of X. Hölder's inequality holds in the form

$$\int_\Omega |f(x)g(x)|\,dx \leq \|f\|_X \|g\|_{X'}$$

for all $f \in X$ and $g \in X'$. Every Banach function space X coincides with its second associate space X'', and $\|f\|_X = \|f\|_{X''}$ for all $f \in X$. In general, the associate space of X is (canonically isomorphic to) a closed subspace of the dual X^*. It is natural to investigate circumstances under which X' coincides with X^*. To do this, we say that $f \in X$ has **absolutely continuous norm** if, for every decreasing sequence $\{G_k\}_{k \in \mathbb{N}}$ of subsets of Ω with $|G_k| \to$

0, we have $\|f\chi_{G_k}\|_X \to 0$; if every $f \in X$ has this property, X is said to have absolutely continuous norm. The following theorem underlines the importance of this notion.

Theorem 1.4. *Let X be a Banach function space. Then*
(i) X' and X^* are canonically isometrically isometric if and only if X has absolutely continuous norm;
(ii) X is reflexive if and only if both X and X' have absolutely continuous norm.

Detailed proofs of these assertions are given in [11], Chapter 1 and [28], Chapter 3.

We now turn to the particular type of Banach function space with which this book is concerned. Let $\mathcal{P}(\Omega)$ be the set of all those measurable functions p on Ω such that $p(\Omega) \subset [1, \infty]$. For each such p put

$$\Omega_0 = \{x \in \Omega : 1 < p(x) < \infty\}, \ \Omega_1 = \{x \in \Omega : p(x) = 1\},$$
$$\Omega_\infty = \{x \in \Omega : p(x) = \infty\},$$

and set

$$p_- = \operatorname*{ess\ inf}_{x \in \Omega_0} p(x) \ \text{and} \ p_+ = \operatorname*{ess\ sup}_{x \in \Omega_0} p(x) \ \text{if} \ |\Omega_0| > 0,$$

$$p_- = p_+ = 1 \ \text{otherwise}.$$

The conjugate function p' is defined by

$$p'(x) = \begin{cases} \dfrac{p(x)}{p(x)-1} & \text{if } x \in \Omega_0, \\ 1 & \text{if } x \in \Omega_\infty, \\ \infty & \text{if } x \in \Omega_1. \end{cases}$$

When greater precision is desirable we shall write Ω_i^p instead of Ω_i. It is easy to check that the function $\rho_{p,\Omega} : \mathcal{M}(\Omega) \to [0, \infty]$ defined by

$$\rho_{p,\Omega}(f) = \int_{\Omega_0 \cup \Omega_1} |f(x)|^{p(x)} \, dx + \operatorname*{ess\ sup}_{x \in \Omega_\infty} |f(x)|$$

is a left-continuous convex modular on $\mathcal{M}(\Omega)$; the corresponding modular space is denoted by $L_{p(\cdot)}(\Omega)$ and is called a **generalised Lebesgue space** or a **Lebesgue space with variable exponent**. Thus

$$L_{p(\cdot)}(\Omega) = \{f \in \mathcal{M}(\Omega) : \rho_{p,\Omega}(\lambda f) < \infty \ \text{for some} \ \lambda > 0\},$$

and the Luxemburg norm on $L_p(\Omega)$ is

$$\|f\|_{p,\Omega} := \inf \{\lambda > 0 : \rho_{p,\Omega}(f/\lambda) \leq 1\}.$$

If the context is clear we shall write ρ_p, $\|\cdot\|_p$ and $L_p(\Omega)$ instead of $\rho_{p,\Omega}$, $\|\cdot\|_{p,\Omega}$ and $L_{p(\cdot)}(\Omega)$. When p is constant, with $p(x) = p$ for all $x \in \Omega$, the space $\left(L_p(\Omega), \|\cdot\|_p \right)$ coincides with the classical Lebesgue space $\left(L_p(\Omega), \|\cdot\|_p \right)$ and the Luxemburg norm equals the classical norm. Note that $L_p(\Omega)$ is an Orlicz-Musielak space with defining function M given by

$$M(x,t) = t^{p(x)}.$$

Proposition 1.4. *Let* $p \in \mathcal{P}(\Omega)$. *Then* $\|f\|_p \leq 1$ *if and only if* $\rho_p(f) \leq 1$; *moreover, if* $\|f\|_p \leq 1$, *then* $\rho_p(f) \leq \|f\|_p$. *If* $f \in L_p(\Omega)\backslash\{0\}$ *and*

$$a := \operatorname*{ess\ sup}_{\{x:f(x)\neq 0\}} p(x) < \infty,$$

then

$$\rho_p\left(f/\|f\|_p\right) = \int_{\Omega_0 \cup \Omega_1} \left(\frac{|f(x)|}{\|f\|_p}\right)^{p(x)} dx + \operatorname*{ess\ sup}_{x \in \Omega_\infty} \frac{|f(x)|}{\|f\|_p} = 1.$$

Proof. The first assertion is contained in Proposition 1.3, the second in Corollary 1.2. Now suppose that $f \in L_p(\Omega)\backslash\{0\}$ and $a < \infty$, assume that $K := \rho_p\left(f/\|f\|_p\right) < 1$ and let $\lambda \in \left(0, \|f\|_p\right)$ be such that $\left(\|f\|_p/\lambda\right)^a K \leq 1$. Then

$$\rho_p(f/\lambda) = \int_{\Omega_0 \cup \Omega_1} \left(\frac{|f(x)|}{\lambda}\right)^{p(x)} dx + \operatorname*{ess\ sup}_{x \in \Omega_\infty} \frac{|f(x)|}{\lambda}$$

$$= \left(\|f\|_p/\lambda\right)^a \left(\lambda/\|f\|_p\right)^a \left(\int_{\Omega_0 \cup \Omega_1} \left(\frac{|f(x)|}{\lambda}\right)^{p(x)} dx + \operatorname*{ess\ sup}_{x \in \Omega_\infty} \frac{|f(x)|}{\lambda}\right)$$

$$\leq \left(\|f\|_p/\lambda\right)^a \left(\int_{\Omega_0 \cup \Omega_1} \left(\lambda/\|f\|_p\right)^a \left(\frac{|f(x)|}{\lambda}\right)^{p(x)} dx + \operatorname*{ess\ sup}_{x \in \Omega_\infty} \frac{|f(x)|}{\|f\|_p}\right)$$

$$\leq \left(\|f\|_p/\lambda\right)^a \left(\int_{\Omega_0 \cup \Omega_1} \left(\frac{|f(x)|}{\|f\|_p}\right)^{p(x)} dx + \operatorname*{ess\ sup}_{x \in \Omega_\infty} \frac{|f(x)|}{\|f\|_p}\right) \leq 1.$$

Since $\lambda < \|f\|_p$, we have a contradiction, and so $K \geq 1$. Let $\{\lambda_k\}_{k \in \mathbb{N}}$ be a decreasing sequence with limit $\|f\|_p$. Then

$$\rho_p(f/\lambda_k) = \int_{\Omega_0 \cup \Omega_1} \left(\frac{|f(x)|}{\lambda_k}\right)^{p(x)} dx + \operatorname*{ess\ sup}_{x \in \Omega_\infty} \frac{|f(x)|}{\lambda_k} \leq 1,$$

so that the integrals $\int_{\Omega_0 \cup \Omega_1} (|f(x)|/\lambda_k)^{p(x)} dx$ are bounded above. Since $|f(x)|/\lambda_k \uparrow |f(x)|/\|f\|_p$, it follows from monotone convergence that

$$K = \int_{\Omega_0 \cup \Omega_1} \left(\frac{|f(x)|}{\|f\|_p}\right)^{p(x)} dx + \operatorname*{ess\,sup}_{x \in \Omega_\infty} \frac{|f(x)|}{\|f\|_p} \leq 1.$$

Thus $K = 1$ and the proof is complete. $\qquad \square$

Sharper relationships between the modular and the norm can be obtained under stricter conditions on p.

Proposition 1.5. *Let $p \in \mathcal{P}(\Omega)$ be such that $1 < p_- \leq p_+ < \infty$. Then for all $f \in L_p(\Omega)$,*

$$\min\left\{\|f\|_p^{p_-}, \|f\|_p^{p_+}\right\} \leq \rho_p(f) \leq \max\left\{\|f\|_p^{p_-}, \|f\|_p^{p_+}\right\}.$$

Proof. First note that by Proposition 1.4, $\rho_p(f/\|f\|_p) = 1$. Thus if $\|f\|_p = \lambda > 1$,

$$\lambda^{-p_+} \rho_p(f) \leq \rho_p(f/\lambda) = 1 \leq \lambda^{-p_-} \rho_p(f),$$

and so

$$\|f\|_p^{p_-} \leq \rho_p(f) \leq \|f\|_p^{p_+}.$$

The case $\|f\|_p < 1$ is handled in a similar manner. $\qquad \square$

Hölder's inequality is valid in the following form.

Theorem 1.5. *If $p \in \mathcal{P}(\Omega)$, then for all $f \in L_p(\Omega)$ and $g \in L_{p'(\cdot)}(\Omega)$,*

$$\int_\Omega |f(x)g(x)|\, dx \leq r_p \|f\|_p \|g\|_{p'},$$

where

$$r_p = c_p + 1/p_- - 1/p_+, \quad c_p = \|\chi_{\Omega_1}\|_\infty + \|\chi_{\Omega_0}\|_\infty + \|\chi_{\Omega_\infty}\|_\infty. \qquad (1.2)$$

Proof. Without loss of generality we suppose that $\|f\|_p \|g\|_{p'} \neq 0$ and $|\Omega_0| > 0$. We use the inequality

$$ab \leq a^q/q + b^{q'}/q'$$

to obtain

$$\int_{\Omega_0} \frac{|f(x)|}{\|f\|_p} \cdot \frac{|g(x)|}{\|g\|_{p'}} dx \leq \|1/p\|_\infty\, \rho_p\left(f/\|f\|_p\right) + \|1/p'\|_\infty\, \rho_{p'}\left(g/\|g\|_{p'}\right)$$

$$\leq 1/p_- + 1 - 1/p^+.$$

Thus

$$\int_\Omega |f(x)g(x)|\, dx \le (1 + 1/p_- - 1/p_+) \|f\|_p \|g\|_{p'} \|\chi_{\Omega_0}\|_\infty$$

$$+ \|f\chi_{\Omega_1}\|_1 \|g\chi_{\Omega_1}\|_\infty + \|f\chi_{\Omega_\infty}\|_\infty \|g\chi_{\Omega_\infty}\|_1$$

$$\le r_p \|f\|_p \|g\|_{p'}. \qquad \square$$

Corollary 1.3. *Suppose that* $p \in \mathcal{P}(\Omega)$. *Then given any* $g \in L_{p'}(\Omega)$, *the functional* G *defined by*

$$G(f) = \int_\Omega f(x)g(x)\, dx \quad (f \in L_p(\Omega))$$

ia a continuous linear functional on $L_p(\Omega)$ *and*

$$c_p^{-1} \|g\|_{p'} \le \|G\| \le r_p \|g\|_{p'}.$$

Proof.   By Proposition 1.3, $\rho_p(f) \le 1$ if and only if $\|f\|_p \le 1$. The result thus follows directly from Theorem 1.5. $\square$

In [55],   Theorem 2.6, it is shown that $p_+ < \infty$ if and only if every continuous linear functional on $L_p(\Omega)$ has a unique representation of this nature.

For functions on Ω we define

$$\|f\|_p' := \sup \left\{ \int f(x)g(x)dx : \rho_{p'}(g) \le 1 \right\}.$$

This is analogous to the Orlicz norm defined earlier for Orlicz spaces. Routine checks show it is a norm on the space of functions f for which $\|f\|_p' < \infty$. In fact, as the next theorem shows, it is a norm equivalent to $\|\cdot\|_p$. Before giving this, however, some preparation is convenient. First we assert that

$$\text{if } |\Omega_1| = |\Omega_\infty| = 0,\ \rho_p(f) < \infty \text{ and } \|f\|_p' \le 1, \text{ then } \rho_p(f) \le 1. \quad (1.3)$$

For suppose that $\rho_p(f) > 1$ and let $\lambda > 1$ be such that $\rho_p(f/\lambda) = 1$; put $g(x) = |f(x)/\lambda|^{p(x)-1} \operatorname{sgn} f(x)$. Then $\rho_{p'}(g) = \rho_p(f/\lambda) = 1$: thus

$$\|f\|_p' \ge \left| \int_\Omega f(x)g(x)\, dx \right| = \lambda \rho_p(f/\lambda) = \lambda > 1.$$

This contradiction establishes the assertion. Next we claim that

$$\text{if } \|f\|_p' \le 1, \text{ then } \rho_p(f) \le c_p \|f\|_p'. \quad (1.4)$$

To begin with, assume that $\rho_p(f) < \infty$, put $f_j = f\chi_{\Omega_j}$ $(j = 0, 1, \infty)$ and note that

$$\rho_p(f) = \|\chi_{\Omega_1}\|_\infty \, \rho_p(f_1) + \|\chi_{\Omega_0}\|_\infty \, \rho_p(f_0) + \|\chi_{\Omega_\infty}\|_\infty \, \rho_p(f_\infty).$$

Set

$$g_1(x) = \mathrm{sgn}\, f_1(x), g_0(x) = |f_0(x)|^{p(x)-1}\, \mathrm{sgn}\, f_0(x).$$

Then $\rho_{p'}(g_1) = \|g_1\|_{\infty,\Omega_1} = 1$ and $\rho_{p'}(g_0) = \int_{\Omega_0} |f(x)|^{p(x)}\, dx \le 1$, by (1.3). Thus

$$\rho_p(f_j) = \int_\Omega f(x)g_j(x)\, dx \le \|f\|'_p \;\; (j = 0, 1).$$

If $|\Omega_\infty| > 0$, then given $\delta \in (0, 1)$, there exists $A \subset \Omega_\infty$ such that $0 < |A| < \infty$ and $|f(x)| \ge \delta$ ess $\sup_{y\in\Omega_\infty} |f(y)|$ $(x \in A)$. Put $g_\infty = |A|^{-1}\chi_A$ sgn f and observe that

$$\rho_{p'}(g_\infty) = \int_A |A|^{-1}\, |\mathrm{sgn}\, f(x)|\, dx \le 1.$$

Hence

$$\|f\|'_p \ge \int_\Omega f(x)g_\infty(x)\, dx$$

$$= \int_A |A|^{-1}\, |f(x)|\, dx \ge \delta \text{ ess } \sup_{y\in\Omega_\infty} |f(y)| = \delta\rho_p(f_\infty).$$

Let $\delta \to 1-$: then $\rho_p(f_\infty) \le \|f\|'_p$ and (1.4) follows, provided that $\rho_p(f) < \infty$. If $\rho_p(f) = \infty$, the result is obtained by using the truncations $f_k := \min\{k, |f|\}\chi_{G_k}$ $(k \in \mathbb{N})$, where the G_k each have finite measure and increase with k, with $\Omega = \cup_{k=1}^\infty G_k$.

Theorem 1.6. *The space $L_p(\Omega)$ coincides with* $\left\{f : \|f\|'_p < \infty\right\}$, *and for all $f \in L_p(\Omega)$,*

$$c_p^{-1}\|f\|_p \le \|f\|'_p \le r_p\|f\|_p,$$

where c_p and r_p are defined in (1.2).

Proof. Suppose first that $f \in L_p(\Omega$. If $\rho_{p'}(g) \le 1$, then $\|g\|_{p'} \le 1$, so that by Theorem 1.5,

$$\left| \int_\Omega |f(x)g(x)|\, dx \right| \le r_p\|f\|_p.$$

Hence $\|f\|_p' < \infty$ and the second of the required inequalities follows.

Now suppose that $0 < \|f\|_p' < \infty$. By (1.4), if $\|h\|_p' \leq 1$, then $\rho_p(h) \leq c_p \|h\|_p'$. Then since $\left\| f/\left(c_p \|f\|_p'\right)\right\|_p' \leq c_p^{-1} \leq 1$, it follows that

$$\rho_p\left(f/\left(c_p \|f\|_p'\right)\right) \leq c_p c_p^{-1} = 1,$$

so that $c_p^{-1}\|f\|_p \leq \|f\|_p'$ and $f \in L_p(\Omega)$. $\square$

The notion of modular convergence is a natural analogue of convergence in the usual norm sense: a sequence $\{f_k\}_{k \in \mathbb{N}}$ in $L_p(\Omega)$ is said to **converge modularly** to $f \in L_p(\Omega)$ if $\lim_{k \to \infty} \rho_p(f - f_k) = 0$. Since $\rho_p(g) \leq \|g\|_p$ whenever $\|g\|_p \leq 1$, norm convergence is stronger than modular convergence. However, we have

Proposition 1.6. *Suppose $p \in \mathcal{P}(\Omega)$ is such that $p_+ < \infty$. Then $\rho_p(f_k) \to 0$ if and only if $\|f_k\|_p \to 0$; moreover, if $\|f_k\|_p \to 0$, then $f_k \to 0$ in measure.*

Proof.   Suppose that $\rho_p(f_k) \to 0$, let $\varepsilon \in (0,1)$ and let $N \in \mathbb{N}$ be such that $\rho_p(f_k) < \varepsilon$ if $k > N$. Then

$$\rho_p\left(\rho_p\left(f_k\right)^{-1/p_+} f_k\right) \leq \rho_p\left(f_k\right)^{-1} \int_{\Omega \backslash \Omega_\infty} |f_k(x)|^{p(x)}\, dx$$

$$+ \rho_p\left(f_k\right)^{-1/p_+} \|f_k\|_{\infty, \Omega_\infty}$$

$$\leq \rho_p\left(f_k\right)^{-1} \rho_p\left(f_k\right) = 1.$$

Thus

$$\|f_k\|_p \leq \rho_p\left(f_k\right)^{1/p_+} < \varepsilon^{1/p^*},$$

so that $\|f_k\|_p \to 0$. That $\|f_k\|_p \to 0$ implies $\rho_p(f_k) \to 0$ is clear.

Finally, suppose that $\|f_k\|_p \to 0$ and f_k does not converge to 0 in measure. Then there exist $\varepsilon > 0$, $\delta \in (0,1)$ and a subsequence $(f_{k_l})_{l \in \mathbb{N}}$ such that for all $l \in \mathbb{N}$,

$$\inf |\{x \in \Omega : |f_{k_l}(x)| > \varepsilon\}| \geq \delta.$$

Thus $\rho_p\left(f_{k_l}\right) \geq \delta \varepsilon^{p_+}$ for all $l \in \mathbb{N}$. But by the first part of the Proposition, $\rho_p(f_{k_l}) \to 0$: contradiction. $\square$

Some basic properties of $L_p(\Omega)$ are given in the following theorem.

Theorem 1.7. *Let $p \in \mathcal{P}(\Omega)$. Then $L_p(\Omega)$ is a Banach function space and its associate space is isomorphic to $L_{p'}(\Omega)$. If $p_+ < \infty$, then the dual of $L_p(\Omega)$ is isomorphic to $L_{p'}(\Omega)$.*

Proof. Properties (i), (ii) and (iv) of a Banach function space are clearly satisfied. For (iii), let $0 \leq f_k \uparrow f$ a.e.: then $\left\{ \|f_k\|_p \right\}$ is non-decreasing. If $\|f\|_p < \infty$, suppose that there exists $\lambda > 0$ such that $\|f_k\|_p \uparrow \lambda < \|f\|_p$. Then

$$1 \geq \int_{\Omega_0 \cup \Omega_1} \left| \frac{f_k(x)}{\|f_k\|_p} \right|^{p(x)} dx + \operatorname*{ess\,sup}_{x \in \Omega_\infty} \left| \frac{f_k(x)}{\|f_k\|_p} \right| \geq \int_{\Omega_0 \cup \Omega_1} \left| \frac{f_k(x)}{\lambda} \right|^{p(x)} dx$$

$$+ \operatorname*{ess\,sup}_{x \in \Omega_\infty} \left| \frac{f_k(x)}{\lambda} \right|$$

$$\uparrow \int_{\Omega_0 \cup \Omega_1} \left| \frac{f(x)}{\lambda} \right|^{p(x)} dx + \operatorname*{ess\,sup}_{x \in \Omega_\infty} \left| \frac{f(x)}{\lambda} \right| > 1,$$

and we have a contradiction. The case in which $\|f\|_p = \infty$ is handled analogously.

To deal with property (iv), let $E \subset \Omega$ be such that $|E| < \infty$, let $f \in X$ and put $M = \{x \in E \cap (\Omega_0 \cup \Omega_1) : |f(x)| < 1\}$, $N = \{x \in E \cap (\Omega_0 \cup \Omega_1) : |f(x)| \geq 1\}$. Then

$$\frac{1}{\|f\|_p} \left(\int_{E \cap (\Omega_0 \cup \Omega_1)} |f(x)|\, dx + \operatorname*{ess\,sup}_{x \in E \cap \Omega_\infty} |f(x)| \right)$$

$$= \int_M \frac{|f(x)|}{\|f\|_p} dx + \int_N \frac{|f(x)|}{\|f\|_p} dx$$

$$+ \operatorname*{ess\,sup}_{x \in E \cap \Omega_\infty} \frac{|f(x)|}{\|f\|_p}$$

$$\leq |M| + \int_N \frac{|f(x)|^{p(x)}}{\|f\|_p^{p(x)}} dx$$

$$+ \operatorname*{ess\,sup}_{x \in E \cap \Omega_\infty} \frac{|f(x)|}{\|f\|_p}$$

$$\leq |M| + 1,$$

and (iv) follows. That the associate space of $L_p(\Omega)$ is isomorphic to $L_{p'}(\Omega)$ follows from the definition of the associate norm and the equivalence of $\|\cdot\|_{p'}$ and $\|\cdot\|'_{p'}$. The remainder of the proof follows as in the proof of Theorem 2.6 of [55]. $\qquad \square$

Note that this result implies that $L_p(\Omega)$ is a Banach space.

Our next objective is to establish, under suitable conditions, various properties of the spaces with variable exponent, such as uniform convexity. We follow the line of reasoning given in [64] and begin with a lemma that characterises spaces with absolutely continuous norms.

Lemma 1.2. *Let $p \in \mathcal{P}(\Omega)$. Then the following statements are equivalent:*

(i) $L_p(\Omega)$ has absolutely continuous norm;
(ii) $p_+ < \infty$;
(iii) $L_p(\Omega)$ is separable.

Proof.   First suppose that $p_+ = \infty$. If $|\Omega_\infty| = 0$, set

$$\Omega^k = \{x \in \Omega : k \leq p(x) < k + 1\} \quad (k \in \mathbb{N})$$

and let $\{n_k\}_{k \in \mathbb{N}}$ be a sequence of natural numbers such that $|\Omega^{n_k}| > 0$ for each $k \in \mathbb{N}$. Let $c_k > 0$ be such that

$$\int\limits_{\Omega^{n_k}} c_k^{p(x)} dx = 1 \quad (k \in \mathbb{N})$$

and put

$$f = \sum_{k=1}^{\infty} c_k \chi_{\Omega^{n_k}}, \quad E_j = \cup_{k=j}^{\infty} \Omega^{n_k};$$

thus $E_j \to \emptyset$. Since

$$\|f\|_p = \inf\left\{\lambda > 0 : \sum_{k=1}^{\infty} \int\limits_{\Omega^{n_k}} \left(\frac{c_k}{\lambda}\right)^{p(x)} dx \leq 1\right\}$$

$$\leq \inf\left\{\lambda > 1 : \sum_{k=1}^{\infty} \left(\frac{1}{\lambda}\right)^{n_k} dx \leq 1\right\} \leq 2,$$

it follows that $f \in L_p(\Omega)$. As

$$\|f\chi_{E_j}\|_p \geq \inf\left\{\lambda > 1 : \int\limits_{\Omega^{n_j}} \left(\frac{c_j}{\lambda}\right)^{p(x)} dx \leq 1\right\} = 1,$$

f does not have absolutely continuous norm. If $|\Omega_\infty| > 0$, take any $A \subset \Omega_\infty$ with positive measure and observe that χ_A evidently does not have absolutely continuous norm. Thus (i) implies (ii).

For the reverse implication, suppose that $p_+ < \infty$, let $f \in L_p(\Omega)$ have norm 1, take $\varepsilon > 0$, let $\{E_k\}_{k \in \mathbb{N}}$ be a sequence of sets with $|E_k| \downarrow 0$, let

$j \in \mathbb{N}$ be such that $\left\| f\chi_{E_j} \right\|_p \geq 1 - \varepsilon$, set $\phi = f\chi_{\Omega \setminus E_j}$ and put $\psi = f\chi_{E_j}$. By Proposition 1.4,

$$\int_{\Omega} \left| \frac{\phi(x)}{\|\phi\|_p} \right|^{p(x)} dx = \int_{\Omega} \left| \frac{\psi(x)}{\|\psi\|_p} \right|^{p(x)} dx = 1.$$

Thus

$$\|\phi\|_p^{p_+} \leq \int_{\Omega} |\phi(x)|^{p(x)} dx, \quad \|\psi\|_p^{p_+} \leq \int_{\Omega} |\psi(x)|^{p(x)} dx$$

and

$$\int_{\Omega} |\phi(x)|^{p(x)} dx + \int_{\Omega} |\psi(x)|^{p(x)} dx \leq 1.$$

Hence

$$\|\psi\|_p^{p_+} \leq \int_{\Omega} |\psi(x)|^{p(x)} dx \leq 1 - \int_{\Omega} |\phi(x)|^{p(x)} dx \leq 1 - \|\phi\|_p^{p_+} \leq 1 - (1-\varepsilon)^{p_+},$$

so that $\|\psi\|_p \leq (1 - (1-\varepsilon)^{p_+})^{1/p_+}$ and the absolute continuity follows. Thus (i) and (ii) are equivalent. For the equivalence of (i) and (iii) we refer to [11], Corollary 1.5.6. $\quad\square$

After this preparation we can give the main result in this area.

Theorem 1.8. *Let $p \in \mathcal{P}(\Omega)$. The following statements are equivalent:*
(i) $L_p(\Omega)$ is reflexive;
(ii) $L_p(\Omega)$ and $L_{p'(\cdot)}(\Omega)$ have absolutely continuous norms;
(iii) $L_p(\Omega)$ is uniformly convex;
(iv) $1 < p_- \leq p_+ < \infty$.

Proof. Since $L_p(\Omega)$ is a Banach function space, the equivalence of (i) and (ii) follows immediately from Theorem 1.4. In view of Lemma 1.2, (ii) implies (iv); and that (iii) implies (i) is obvious. All that is left is to show that (iv) implies (iii). Suppose that (iv) holds, let $\varepsilon \in (0,1)$ and let u, v have unit $L_p(\Omega)$ norm. Set $s = (u+v)/2, t = (u-v)/2$ and $\Gamma = \{x \in \Omega : p_- \leq p(x) \leq p_+\}$, so that $|\Omega \setminus \Gamma| = 0$; put

$$S = \{x \in \Gamma : |t(x)| < \varepsilon |s(x)|\}, T = \{x \in \Gamma : |t(x)| \geq \varepsilon |s(x)|\}.$$

Then

$$\int_S |t(x)|^{p(x)} dx \leq \int_S \varepsilon^{p(x)} |s(x)|^{p(x)} dx \leq \int_{\Omega} \varepsilon^{p(x)} |s(x)|^{p(x)} dx \qquad (1.5)$$

$$\leq \varepsilon^{p_-} \int_{\Omega} |s(x)|^{p(x)} dx \leq \varepsilon^{p_-}.$$

The function $\lambda \longmapsto \lambda^t$ is strictly convex on $\mathbb{R}$ whenever $1 < t < \infty$; hence

$$|\lambda|^t < \left(|\lambda + 1|^t + |\lambda - 1|^t\right)/2 \ (\lambda \in \mathbb{R}). \tag{1.6}$$

Thus the function $f \ : \ (q, \lambda) \ \longmapsto \ \left(|\lambda + 1|^q + |\lambda - 1|^q\right)/2 - |\lambda|^q$ $(q \in (1, \infty), \lambda \in \mathbb{R})$ is continuous and strictly positive on $(1, \infty) \times \mathbb{R}$. It follows that there exists $\alpha > 0$ such that $f(q, \lambda) \geq \alpha$ whenever $q \in [p_-, p_+]$ and $\lambda \in [-1/\varepsilon, 1/\varepsilon]$. We therefore see that

$$\frac{1}{2}\left(|\lambda + 1|^{p(x)} + |\lambda - 1|^{p(x)}\right) - |\lambda|^{p(x)} \geq \alpha$$

for all $x \in \Gamma$ and $\lambda \in [-1/\varepsilon, 1/\varepsilon]$, so that

$$\frac{1}{2}\left(|s(x) + t(x)|^{p(x)} + |s(x) - t(x)|^{p(x)}\right) \geq |s(x)|^{p(x)} + \alpha\,|t(x)|^{p(x)} \quad (x \in T),$$

while by (1.6),

$$\frac{1}{2}\left(|s(x) + t(x)|^{p(x)} + |s(x) - t(x)|^{p(x)}\right) \geq |s(x)|^{p(x)} \quad (x \in S).$$

Thus

$$1 = \int_\Omega \frac{1}{2}\left(|s(x) + t(x)|^{p(x)} + |s(x) - t(x)|^{p(x)}\right) dx$$

$$\geq \int_\Omega |s(x)|^{p(x)}\, dx + \alpha \int_T |t(x)|^{p(x)}\, dx,$$

and so

$$\int_T |t(x)|^{p(x)}\, dx < \varepsilon^{p_-} \text{ if } \int_\Omega |s(x)|^{p(x)}\, dx > 1 - \alpha\varepsilon^{p_-}.$$

Thus if $\delta := \alpha\varepsilon^{p_-}$ and $\|(u + v)/2\| > 1 - \delta$, then $\int_\Omega |s(x)|^{p(x)}\, dx > 1 - \delta$, from which with the aid of (1.5) it follows that

$$\int_\Omega |t(x)|^{p(x)}\, dx < 2\varepsilon^{p_-},$$

so that

$$\|u - v\| = \|2t\| \leq 2(2\varepsilon^{p_-})^{1/p_+},$$

which establishes the uniform convexity of $L_p(\Omega)$. $\qquad\square$

This theorem shows that if $1 < p_- \leq p_+ < \infty$, the space $L_p(\Omega)$ has very good properties. Interesting results can still be obtained under weaker hypotheses, as we now show, following [55].

Theorem 1.9. *Let $p \in \mathcal{P}(\Omega)$ and suppose that $p_+ < \infty$. Then the set of all bounded, measurable functions is dense in $L_p(\Omega)$, as is $C(\Omega) \cap L_p(\Omega)$; and $L_p(\Omega)$ is separable. If Ω is open, then $C_0^\infty(\Omega)$ is dense in $L_p(\Omega)$.*

Proof. Let $f \in L_{p)}(\Omega)$, and for each $k \in \mathbb{N}$ let $G_k = \{x \in \Omega \backslash \Omega_\infty : |x| < k\}$ and

$$f_k(x) = \begin{cases} f(x), & \text{if } |f(x)| \leq k \text{ and } x \in G_k \cup \Omega_\infty, \\ k \, \text{sgn} \, f(x), & \text{if } |f(x)| > k \text{ and } x \in G_k \cup \Omega_\infty, \\ 0 & \text{at} \qquad \text{all other points of } \Omega. \end{cases}$$

The measurable functions f_k are bounded on Ω and by dominated convergence, $\rho_p(f - f_k) \to 0$ as $k \to \infty$. By Proposition 1.6, $\|f - f_k\|_p \to 0$. This establishes the density in $L_p(\Omega)$ of the set of all bounded measurable functions.

Next, let $f \in L_p(\Omega)$ and $\varepsilon > 0$. In view of what has just been proved, there exists a bounded function g such that $\|f - g\|_p < \varepsilon$. By Lusin's theorem (see, for example, [56], p.63), there exist $h \in C(\Omega)$ and an open set U with $|U| < \min \{1, (\varepsilon/(2 \|g\|_\infty))^{p_+}\}$, such that

$$g(x) = h(x) \text{ if } x \in \Omega \backslash U, \quad \sup_{x \in \Omega} |h(x)| = \sup_{x \in \Omega \backslash U} |g(x)| \leq \|g\|_\infty.$$

Thus

$$\rho_p((g - h)/\varepsilon) \leq |U| \max \{1, (2 \|g\|_\infty /\varepsilon)^{p_+}\}.$$

Hence $\|g - h\|_p \leq \varepsilon$, so that $\|f - h\|_p \leq 2\varepsilon$.

Finally, suppose that Ω is open. Since $p \in L_\infty(\Omega)$, it follows that $C_0^\infty(\Omega) \subset L_p(\Omega)$ and $\rho_p(h/\varepsilon) < \infty$, so that there exists a bounded open set $G \subset \Omega$ such that $\rho_p(h\chi_{\Omega \backslash G}/\varepsilon) \leq 1$. Thus $\|h - h\chi_G\|_p \leq \varepsilon$. Let m be a polynomial such that $\sup_G |h(x) - m(x)| < \varepsilon \min \{1, |G|^{-1}\}$. Then

$$\rho_p((h\chi_G - m\chi_G)/\varepsilon) \leq |G| \min \{1, |G|^{-1}\} \leq 1,$$

which implies that $\|h\chi_G - m\chi_G\|_p \leq \varepsilon$. In the same way it follows that for sufficently small $a > 0$ the compact set $K_a := \{x \in G : \text{dist}\,(x, \partial G) \geq a\}$ satisfies $\|m\chi_G - m\chi_{K_a}\|_p \leq \varepsilon$. Let $\phi \in C_0^\infty(G)$ be such that $0 \leq \phi(x) \leq 1$ for all $x \in G$, $\phi(x) = 1$ if $x \in K_a$: then

$$\|m\chi_G - m\phi\|_p \leq \|m\chi_G - m\chi_{K_a}\|_p \leq \varepsilon,$$

and finally we have $\|f - m\phi\|_p < 4\varepsilon$.

To establish separability, for each $k \in \mathbb{N}$ let Ω_k be a bounded subset of Ω such that $\Omega_k \subset \Omega_{k+1}$ and $\Omega = \cup_{k=1}^{\infty}\Omega_k$. As above it can be shown that the countable set

$$\{P\chi_{\Omega_k} : k \in \mathbb{N}, \ P \text{ is a polynomial on } \mathbb{R}^n \text{ with rational coefficients}\}$$

is dense in $L_p(\Omega)$, which completes the proof. $\square$

Note that if the hypothesis that $p_+ < \infty$ is removed, $C_0^\infty(\Omega)$ need not have the density property. For example, let $n = 1, \Omega = (1, \infty)$ and $p(x) = x$ $(x \in \Omega)$; let f be the constant function on Ω that is everywhere equal to 1. Then $f \in L_p(\Omega)$ since

$$\int\limits_{1}^{\infty} \lambda^{-x}dx = (\lambda \log \lambda)^{-1} \text{ if } \lambda > 1.$$

Let $g \in C_0^\infty(\Omega)$, so that supp $g \subset [1 + \varepsilon, N]$ for some $\varepsilon > 0$ and $N > 1$. Hence

$$\|f - g\|_p \geq \inf \left\{ \lambda > 0 : \int\limits_{N}^{\infty} \lambda^{-x}dx \leq 1 \right\},$$

and as the integral is finite only if $\lambda > 1$, it follows that $\|f - g\|_p \geq 1$. Thus $C_0^\infty(\Omega)$ is not dense in $L_p(\Omega)$.

Let X be a Banach function space. We distinguish three subspaces of X :

(i) X_a, the set of all elements of X with absolutely continuous norm;

(ii) X_b, the closure of the set of all bounded functions supported in sets of finite measure;

(iii) X_c, the space of all functions f with a continuous norm, by which we mean that for every $x \in \overline{\Omega}$ we have

$$\lim_{\varepsilon \to 0} \left\| f\chi_{B(x,\varepsilon)\cap\Omega} \right\|_p = 0.$$

From [11] we have that X_a is a closed subspace of X and $X_a \subset X_b$; from [59] we know that X_c is a closed subspace of X and $X_a \subset X_c$. Now take $X = L_p(\Omega)$. Then in view of Lemma 1.2 it follows that

$$\text{if } p_+ < \infty, \text{ then } X_a = X_b = X_c = X. \tag{1.7}$$

The position when p_+ may be infinite is more complicated, as we now illustrate with the next two theorems from [32].

Theorem 1.10. *Suppose that Ω is bounded and $p \in \mathcal{P}(\Omega)$. Then $X_a = X_c$.*

Proof. We simply have to show that $X_c \subset X_a$. First suppose that $|\Omega_\infty| = 0$. Assume that $f \notin X_a$. Then there exist $\alpha > 0$ and a decreasing sequence of sets $\{G_k\}_{k \in \mathbb{N}}$ such that $|G_k| \to 0$ and, for every $k \in \mathbb{N}$,

$$\|f\chi_{G_k}\|_p = \inf \left\{ \lambda > 0 : \int\limits_{G_k} (f(x)/\lambda)^{p(x)} \, dx \leq 1 \right\} \geq \alpha.$$

With $\beta := \alpha/2$ this implies that

$$\int\limits_{G_k} (f(x)/\beta)^{p(x)} \, dx > 1 \ (k \in \mathbb{N}).$$

If there existed m such that

$$\int\limits_{G_m} (f(x)/\beta)^{p(x)} \, dx \leq K < \infty,$$

then as the G_k are decreasing, with $|G_k| \to 0$, by dominated convergence we would have

$$\lim_{k \to \infty} \int\limits_{G_k} (f(x)/\beta)^{p(x)} \, dx = 0,$$

a contradiction. Thus

$$\int\limits_{G_k} (f(x)/\beta)^{p(x)} \, dx = \infty$$

for every $k \in \mathbb{N}$.

Since Ω is bounded, there is a closed cube Q with side length $l(Q)$ such that $\Omega \subset Q$. By subdivision of Q we obtain a decreasing sequence $\{Q_k\}_{k \in \mathbb{N}}$ of closed cubes with $l(Q_k) = 2^{-k}l(Q)$ and

$$\int\limits_{Q_k} (f(x)/\beta)^{p(x)} \, dx = \infty$$

for every k. Let $x \in \cap_{k=1}^{\infty} Q_k$ and $\varepsilon > 0$. Then $Q_k \subset B(x, \varepsilon)$ for large enough k and so

$$\int\limits_{\Omega \cap B(x,\varepsilon)} (f(x)/\beta)^{p(x)} \, dx \geq \int\limits_{\Omega \cap Q_k} (f(x)/\beta)^{p(x)} \, dx$$

$$= \int\limits_{\Omega} (f(x)\chi_{Q_k}(x)/\beta)^{p(x)} \, dx = \infty.$$

Hence $\left\|f\chi_{\Omega\cap B(x,\varepsilon)}\right\|_p > \beta$, and so $f \notin X_c$. This establishes the theorem when $|\Omega_\infty| = 0$.

Finally, we dispense with this assumption and take any $f \in X_c$. The norms $\|f\|_p$ and $\left\|f\chi_{\Omega\setminus\Omega_\infty}\right\|_p + \text{ess sup}_{x\in\Omega_\infty} |f(x)|$ are equivalent. Since $f \in X_c$, $f(x) = 0$ on Ω_∞. Now apply the case of the theorem already proved to the set $\Omega\setminus\Omega_\infty$ and the function $p_1 := p\chi_{\Omega_1}$ to obtain $f \in X_a$. $\square$

Theorem 1.11. *Let $p \in \mathcal{P}(\Omega)$ and suppose that Ω is bounded; for each $k \in \mathbb{N}$ put $\Lambda_k = \{x \in \Omega : k \leq p(x) < k + 1\}$.*
(i) If $p \in L_\infty(\Omega\setminus\Omega_\infty)$, then $X_b = X$;
(ii) if $|\Omega_\infty| = 0$ and $p_+ = \infty$, then $X_b \neq X$;
(iii) if $|\Omega_\infty| = 0$ and $p_+ = \infty$, then $X_a = X_b$ if and only if

$$\sum_{k=1}^{\infty} A^k |\Lambda_k| < \infty \text{ for all } A > 1. \tag{1.8}$$

Proof.   (i) This is immediate from (1.7) and the fact that $L_\infty(\Omega_\infty) = X_b(\Omega_\infty)$.

(ii) For each $k \in \mathbb{N}$ let c_k be chosen so that $\int_{\Omega_k} c_k^{p(x)} dx = 2^{-k}$ if $|\Lambda_k| > 0$, $c_k = 0$ otherwise; put $d_k = \max(c_k, 2k)$ if $c_k > 0$, $d_k = 0$ otherwise. There are sets $\Lambda_k' \subset \Lambda_k$ such that

$$\int_{\Lambda_k'} d_k^{p(x)} dx = 2^{-k} \text{ if } |\Lambda_k| > 0.$$

Let $f = \sum_{k=1}^{\infty} d_k \chi_{\Lambda_k'}$. Since $f(x) = 0$ on Ω_∞,

$$\rho_p(f) = \int_{\Omega\setminus\Omega_\infty} f(x)^{p(x)} dx = \sum_{k=1}^{\infty} \int_{\Lambda_k'} d_k^{p(x)} dx \leq \sum_{k=1}^{\infty} 2^{-k} = 1;$$

hence $\|f\|_p \leq 1$.

Now define

$$f_k(x) = \begin{cases} f(x), & 0 \leq f(x) \leq k, \\ k, & f(x) > k, \end{cases}$$

and let g be any measurable function on Ω with $|g| \leq k$. Since $|f(x) - g(x)| \geq |f(x) - f_k(x)|$ we have $\|f - g\|_p \geq \|f - f_k\|_p$; moreover,

$f - f_k = 0$ on Ω_∞. Thus

$$\int_{\Omega\backslash\Omega_\infty} (4\,|f(x) - f_k(x)|)^{p(x)}\,dx \geq \sum_{j=k}^{\infty} 4^j \int_{\Lambda_j'} (d_j - k)^{p(x)}dx$$

$$\geq \sum_{j=k}^{\infty} 4^j \int_{\Lambda_j'} (d_j/2)^{p(x)}dx \geq \sum_{j=k}^{\infty} \frac{4^j}{2^{j+1}} \int_{\Lambda_j'} d_j^{\,p(x)}dx.$$

But $\int_{\Lambda_j'} d_j^{\,p(x)}dx = 2^{-j}$ for infinitely many j : hence the last sum is infinite and $\|f - g\|_p > 1/4$. Since g is an arbitrary bounded function, the proof of (ii) is complete.

(iii) Suppose that (1.8) holds and let f be a bounded function on Ω with $|f| \leq K$. Let $\{G_m\}_{m\in\mathbb{N}}$ be a decreasing sequence of measurable subsets of Ω with $|G_m| \to 0$, and let $\lambda \in (0, K)$. Then for every m,

$$\int_{G_m} \left|\frac{f(x)}{\lambda}\right|^{p(x)} dx \leq \int_{G_m} \left(\frac{K}{\lambda}\right)^{p(x)} dx = \sum_{k=1}^{\infty} \int_{G_m \cap \Lambda_k} \left(\frac{K}{\lambda}\right)^{p(x)} dx$$

$$\leq \sum_{k=1}^{\infty} \left(\frac{K}{\lambda}\right)^{k+1} |G_m \cap \Lambda_k|.$$

Since $K\backslash\lambda > 1$, (1.8) implies that there exists k_0 such that

$$\sum_{k=k_0+1}^{\infty} \left(\frac{K}{\lambda}\right)^{k+1} |\Lambda_k| \leq \frac{1}{2};$$

as $|G_m| \to 0$, there exists m_0 such that if $m \geq m_0$, then

$$\sum_{k=1}^{\infty} \left(\frac{K}{\lambda}\right)^{k+1} |G_m \cap \Lambda_k| \leq \frac{1}{2}.$$

Thus

$$\int_{G_m} \left|\frac{f(x)}{\lambda}\right|^{p(x)} dx \leq 1 \text{ if } m \geq n_0.$$

Thus for any small $\lambda > 0$ we have $\|f\chi_{G_m}\|_p \leq \lambda$ if $m \geq m_0$, so that $\lim_{m\to\infty} \|f\chi_{G_m}\|_p = 0$. Hence $f \in X_a$.

Now assume that $f \in X_b$. Then there is a sequence of bounded functions f_m with $\|f - f_m\|_p \to 0$. By the last part of the proof, each $f_m \in X_a$; as X_a is closed, $f \in X_a$. Now suppose that there exists $A > 1$ such that

$\sum\limits_{k=1}^{\infty} A^k |\Lambda_k| = \infty$. Let g be the function that is identically 1 on Ω, and put $G_m = \cup_{k=m}^{\infty}\Lambda_k$. Then

$$\int_{\Omega} \left(\frac{f\chi_{G_m}}{1/A}\right)^{p(x)} dx = \sum_{k=1}^{\infty}\int_{\Lambda_k} A^{p(x)}\chi_{G_m}(x)dx \geq \sum_{k=m}^{\infty} A^m |\Lambda_k| = \infty.$$

Hence $\|f\chi_{G_m}\|_p > 1/A$ for any m. As the G_m are decreasing and $|G_m| \to 0$, it follows that $f \notin X_a$. Since plainly $f \in X_b$ the proof is complete. $\qquad\square$

Note that under the conditions (iii) of the above theorem another characteristion of the equality of X_a and X_b can be given in terms of the non-increasing rearrangement p^* of the function p. In fact, since

$$\int_0^{|\Omega|} A^{p^*(t)} dt = \int_{\Omega} A^{p(x)} dx,$$

it is immediate that $X_a = X_b$ if and only if

$$\int_0^{|\Omega|} A^{p^*(t)} dt < \infty \quad \text{for all } A > 1.$$

By way of illustration, consider the case in which $n = 1$, $\Omega = (0, 1/e)$ and $p^*(x) = x^\alpha$ for some $\alpha < 0$. Then

$$\int_0^{1/e} A^{p^*(x)} dx = \int_0^{1/e} A^{x^\alpha} dx = \int_0^{1/e} \sum_{k=0}^{\infty} \frac{(x^\alpha \log A)^k}{k!} dx$$
$$= \infty$$

for all $A > 1$. Thus $X_a = X_b$.

On the other hand, if $n = 1$, $\Omega = (0, 1/e)$ and $p^*(x) = (\log x^{-1})^\alpha$ for some $\alpha \geq 0$, then

$$\int_0^{1/e} A^{p^*(x)} dx = \int_0^{1/e} \exp\left((\log x^{-1})^\alpha \log A\right) dx = \int_1^{\infty} \exp\left(y^\alpha \log A - y\right) dy$$
$$< \infty$$

for all $\alpha \in (0, 1)$ and all $A > 1$: thus $X_a = X_b$ when $\alpha \in (0, 1)$. However, if $\alpha \in [1, \infty)$, the choice $A = e$ gives

$$\int_0^{1/e} A^{p^*(x)} dx = \int_1^{\infty} \exp\left(y^\alpha \log A - y\right) dy = \infty,$$

and so $X_a \subsetneq X_b$.

Next we consider embeddings. It is well known that when Ω has finite measure, the classical Lebesgue spaces are ordered: if $p < q$, then $L_q(\Omega) \hookrightarrow$

$L_p(\Omega)$. That the same holds for spaces with variable exponent was first shown by Kováčik and Rákosník in Theorem 2.8 of [21]: their argument is given below.

Theorem 1.12. *Let $|\Omega| < \infty$ and suppose that $p, q \in \mathcal{P}(\Omega)$. Then $L_q(\Omega)$ is continuously embedded in $L_p(\Omega)$ if and only if $p(x) \leq q(x)$ a.e. in Ω. When the embedding id exists,*

$$\|id\| \leq 1 + |\Omega|.$$

Proof. First suppose that $p(x) \leq q(x)$ a.e. in Ω. Note that $\Omega_\infty^p \subset \Omega_\infty^q$. Let f belong to the closed unit ball of $L_q(\Omega)$. Then by Proposition 1.4,

$$\rho_q(f) = \int\limits_{\Omega\backslash\Omega_\infty^q} |f(x)|^{q(x)}\, dx + \text{ess sup}_{\Omega_\infty^q} |f(x)| \leq 1.$$

Thus $|f(x)| \leq 1$ a.e. in Ω_∞^q. Hence

$$\rho_p(f) \leq |\{x \in \Omega\backslash\Omega_\infty^q : |f(x)| \leq 1\}|$$
$$+ \int\limits_{\Omega\backslash\Omega_\infty^q} |f(x)|^{q(x)}\, dx + |\Omega_\infty^q \backslash \Omega_\infty^p| + \text{ess sup}_{\Omega_\infty^q} |f(x)|$$
$$\leq |\Omega| + \rho_q(f) \leq |\Omega| + 1.$$

As ρ_p is convex,

$$\rho_p\left(f/(|\Omega| + 1)\right) \leq (|\Omega| + 1)^{-1} \rho_p(f) \leq 1,$$

from which we see that $\|id\| \leq 1 + |\Omega|$.

Now suppose that it is not the case that $p(x) \leq q(x)$ a.e. in Ω. Then there is a subset Ω^* of Ω, with positive measure, such that $p(x) > q(x)$ for all $x \in \Omega^*$. First assume that $|\Omega_\infty^p \cap \Omega^*| > 0$. Then there exist $A \subset \Omega_\infty^p \cap \Omega^*$, with $0 < |A| < \infty$, and $r \in (1, \infty)$ such that $1 \leq q(x) \leq r < \infty = p(x)$ for all $x \in A$. There are pairwise disjoint sets A_k such that $A = \cup_{k=1}^\infty A_k$ and $|A_k| = 2^{-k}|A|$ for all $k \in \mathbb{N}$; define $f = \sum\limits_{k=1}^\infty (3/2)^r \chi_{A_k}$. For this function f we have $\|f\|_p \geq \|f\chi_A\|_\infty = \infty$, so that $f \notin L_p(\Omega)$. However,

$$\rho_q(f) = \int\limits_A |f(x)|^{q(x)}\, dx = \sum_{k=1}^\infty \int\limits_{A_k} (3/2)^{kq(x)/r}\, dx$$
$$\leq \sum_{k=1}^\infty (3/2)^k |A_k| = |A| \sum_{k=1}^\infty (3/4)^k < \infty,$$

and hence $f \in L_q(\Omega)$.

On the other hand, if $|\Omega_\infty^p \cap \Omega^*| = 0$, then $1 \leq q(x) < p(x) < \infty$ for a.e. $x \in \Omega^*$ and there is a set $A \subset \Omega^*$, with $0 < |A| < \infty$, and numbers $a > 0, r \in (1, \infty)$ such that $q(x) + a \leq p(x) \leq r$ whenever $x \in A$. Choose sets A_k as before and define $f(x) = \sum_{k=1}^{\infty} (2^k k^{-2})^{1/q(x)}$ $(x \in \Omega)$. Then

$$\rho_q(f) = \sum_{k=1}^{\infty} 2^k k^{-2} |A_k| = |A| \sum_{k=1}^{\infty} k^{-2} < \infty,$$

which shows that $f \in L_q(\Omega)$. But since for all $\lambda \in (0, 1]$,

$$\rho_p(\lambda f) \geq \lambda^r \sum_{k=1}^{\infty} \int_{A_k} (2^k k^{-2})^{p(x)/q(x)} dx \geq \lambda^r \sum_{k=1}^{\infty} (2^k k^{-2})^{1+a/r} |A_k|$$

$$= \lambda^r |A| \sum_{k=1}^{\infty} 2^{ak/r} k^{-2(1+a/r)} = \infty,$$

we see that $f \notin L_p(\Omega)$. $\qquad\qquad\qquad\qquad\qquad\qquad\qquad\qquad\square$

Corollary 1.4. *Suppose that $p \in \mathcal{P}(\Omega)$ and let $(f_k)_{k \in \mathbb{N}}$ be a sequence in $L_p(\Omega)$ that converges to f in $L_p(\Omega)$. Then there is a subsequence of $(f_k)_{k \in \mathbb{N}}$ that converges pointwise a.e. in Ω to f.*

Proof. First suppose that $|\Omega| < \infty$. By Theorem 1.12, $L_p(\Omega) \hookrightarrow L_1(\Omega)$ and so $\|f - f_k\|_1 \to 0$. As the result is true for $L_1(\Omega)$, there is nothing more to do. If $|\Omega| = \infty$, apply what has already been proved to $L_p(\Omega \cap B(0, m))$ $(m \in \mathbb{N})$ and use diagonalisation. $\qquad\qquad\qquad\qquad\qquad\qquad\square$

Corollary 1.5. *Let $p \in \mathcal{P}(\Omega)$ be such that $p_+ < \infty$ and let $f, f_k \in L_p(\Omega)$ $(k \in \mathbb{N})$. Then $f_k \to f$ in $L_p(\Omega)$ if and only if*

$$(f_k) \text{ converges to } f \text{ in measure on } \Omega \text{ and } \rho_p(f_k) \to \rho_p(f).$$

Proof. Suppose that $f_k \to f$ in $L_p(\Omega)$. By Proposition 1.6, $(f_k)_{k \in \mathbb{N}}$ converges to f in measure on Ω and $\rho_p(f_k - f) \to 0$. By Corollary 1.4, there is a subsequence $(f_{l(k)})_{k \in \mathbb{N}}$ of $(f_k)_{k \in \mathbb{N}}$ that converges pointwise a.e. on Ω to f. The inequality

$$|f_k(x)|^{p(x)} \leq 2^{p_+ - 1} \left(|f_k(x) - f(x)|^{p(x)} + |f(x)|^{p(x)} \right)$$

shows that given $\varepsilon > 0$, there exist $\delta > 0$ and $k_0 \in \mathbb{N}$ such that for all $B \subset \Omega$ with $|B| < \delta$ and all $k \geq k_0$,

$$\int_B |f_k(x)|^{p(x)} \, dx \leq \varepsilon.$$

Hence by Vitali's theorem (see [56], 2.1.4),

$$\lim_{k \to \infty} \rho_p(f_{l(k)}) = \rho_p(f).$$

In fact, $\lim_{k \to \infty} \rho_p(f_k) = \rho_p(f)$; for if not, there would be a subsequence of $(\rho_p(f_k))_{k \in \mathbb{N}}$, the elements of which were uniformly bounded away from $\rho_p(f)$, and application of the preceding argument to this subsequence would give a contradiction.

For the converse, suppose that $(f_k)_{k \in \mathbb{N}}$ converges to f in measure on Ω and $\rho_p(f_k) \to \rho_p(f)$. Now use the inequality

$$|f_k(x) - f(x)|^{p(x)} \leq 2^{p_+ - 1} \left(|f_k(x)|^{p(x)} + |f(x)|^{p(x)} \right),$$

and argue as in the first part to obtain $\rho_p(f_k - f) \to 0$ and hence $\|f_k - f\|_p \to 0$. $\qquad\square$

In addition to the convergence results of the last two corollaries, it can be shown (see, for example, [23], Lemma 2.12 and Theorem 2.13) that:
(a) if $f_k \to f$ a.e. in Ω, then $\rho_p(f) \leq \liminf_{k \to \infty} \rho_p(f_k)$;
(b) if $0 \leq f_k \uparrow f$ a.e. in Ω, then $\rho_p(f) \leq \liminf_{k \to \infty} \rho_p(f_k)$ (the Fatou property of the modular);
(c) if $f_k \rightharpoonup f$ in L_p, then $\rho_p(f) \leq \liminf_{k \to \infty} \rho_p(f_k)$.

We now show (following [33]) that for embedding maps, greater precision in norm estimation can be obtained when p and q are near to one another in the sense that for some $\varepsilon \in (0, 1)$,

$$p(x) \leq q(x) \leq p(x) + \varepsilon \text{ a.e. in } \Omega. \tag{1.9}$$

Lemma 1.3. *Suppose that $|\Omega| < \infty$ and p, q satisfy (1.9). If $f \in \mathcal{M}(\Omega)$ and $\rho_q(f) \leq 1$, then*

$$\rho_p(f) \leq \varepsilon |\Omega| + \varepsilon^{-\varepsilon}.$$

Proof. Put

$$\Omega^1 = \{x \in \Omega : |f(x)| < \varepsilon\}, \Omega^2 = \{x \in \Omega : \varepsilon \leq |f(x)| \leq 1\},$$
$$\Omega^3 = \{x \in \Omega : 1 < |f(x)|\}.$$

Then

$$\rho_p(f) = \sum_{j=1}^{3} \int_{\Omega^j} |f(x)|^{p(x)} \, dx = \sum_{j=1}^{3} A_j, \text{ say.}$$

It is clear that

$$A_1 \leq \int_{\Omega^1} \varepsilon^{p(x)} dx \leq \int_{\Omega^1} \varepsilon dx \leq \varepsilon |\Omega|, \quad A_3 \leq \int_{\Omega^3} |f(x)|^{q(x)} \, dx.$$

On Ω^2,

$$\varepsilon^\varepsilon \leq \varepsilon^{q(x)-p(x)} \leq |f(x)|^{q(x)-p(x)} \leq 1,$$

and so

$$1 \leq |f(x)|^{p(x)-q(x)} \leq \varepsilon^{-\varepsilon}.$$

Hence

$$A_2 = \int_{\Omega^2} |f(x)|^{q(x)} |f(x)|^{p(x)-q(x)} \, dx \leq \varepsilon^{-\varepsilon} \int_{\Omega^2} |f(x)|^{q(x)} \, dx.$$

Thus

$$\rho_p(f) \leq \varepsilon |\Omega| + \varepsilon^{-\varepsilon} \int_{\Omega^2} |f(x)|^{q(x)} \, dx + \int_{\Omega^3} |f(x)|^{q(x)} \, dx$$

$$\leq \varepsilon |\Omega| + \varepsilon^{-\varepsilon} \left(\int_{\Omega^2} |f(x)|^{q(x)} \, dx + \int_{\Omega^3} |f(x)|^{q(x)} \, dx \right)$$

$$\leq \varepsilon |\Omega| + \varepsilon^{-\varepsilon} \int_{\Omega} |f(x)|^{q(x)} \, dx \leq \varepsilon |\Omega| + \varepsilon^{-\varepsilon}.$$

$\square$

Lemma 1.4. *Suppose that $|\Omega| < \infty$ and p, q satisfy (1.9). Then*

$$\|id\| \leq \varepsilon |\Omega| + \varepsilon^{-\varepsilon}.$$

Proof. Clearly $K := \varepsilon |\Omega| + \varepsilon^{-\varepsilon} > 1$. For each f such that $\rho_q(f) \leq 1$ we have, by Lemma 1.3,

$$\rho_p(f/K) \leq K^{-1} \rho_p(f) \leq \left(\varepsilon |\Omega| + \varepsilon^{-\varepsilon} \right) / K = 1,$$

and the result follows. $\square$

The required estimate from below is established in two stages.

Lemma 1.5. *Suppose that $1 \leq |\Omega| < \infty$ and p, q satisfy (1.9). Then*

$$\|id\| \geq 1.$$

Proof. Let $g(x) = |\Omega|^{-1/q(x)}$ $(x \in \Omega)$. Then $\rho_q(g) = 1$. Since $|\Omega|^{-p(x)/q(x)} \geq |\Omega|^{-1}$ we have, for each $\lambda \in (0,1)$,

$$\rho_p(g/\lambda) = \int_\Omega \frac{|\Omega|^{-p(x)/q(x)}}{\lambda^{p(x)}} dx \geq \int_\Omega \frac{|\Omega|^{-1}}{\lambda^{p(x)}} dx \geq \int_\Omega \frac{|\Omega|^{-1}}{\lambda} dx = \lambda^{-1} > 1.$$

Hence $\|\mathrm{id}\| \geq \lambda$ for each $\lambda \in (0,1)$, which shows that $\|\mathrm{id}\| \geq 1$. $\qquad\square$

Lemma 1.6. *Suppose that $0 < |\Omega| < 1$ and p, q satisfy (1.9). Then $\|\mathrm{id}\| \geq |\Omega|^\varepsilon$.*

Proof. Let g be as in the last lemma. Since

$$|\Omega|^{1-p(x)/q(x)} \geq |\Omega|^{\varepsilon/q(x)} \geq |\Omega|^\varepsilon,$$

we see that

$$\rho_p(g) = \int_\Omega |\Omega|^{-p(x)/q(x)} dx = |\Omega|^{-1} \int_\Omega |\Omega|^{1-p(x)/q(x)} dx \geq |\Omega|^\varepsilon.$$

Thus for each positive $\lambda < |\Omega|^\varepsilon$,

$$\rho_p(g/\lambda) > \int_\Omega \left| \frac{g(x)}{|\Omega|^\varepsilon} \right|^{p(x)} dx = \int_\Omega \left| \frac{g(x)}{|\Omega|^{\varepsilon/p(x)}} \right|^{p(x)} dx$$

$$= |\Omega|^{-\varepsilon} \int_\Omega |g(x)|^{p(x)} dx \geq |\Omega|^{-\varepsilon} |\Omega|^\varepsilon = 1.$$

Hence $\|\mathrm{id}\| \geq \lambda$ for each positive $\lambda < |\Omega|^\varepsilon$, and the result follows. $\qquad\square$

From these lemmas we immediately have the following theorem and corollary.

Theorem 1.13. *Suppose that $0 < |\Omega| < \infty$ and p, q satisfy (1.9). Then the norm of the embedding id of $L_q(\Omega)$ in $L_p(\Omega)$ satisifes*

$$\min(1, |\Omega|^\varepsilon) \leq \|\mathrm{id}\| \leq \varepsilon |\Omega| + \varepsilon^\varepsilon.$$

Corollary 1.6. *Suppose that $0 < |\Omega| < \infty$ and $p \in \mathcal{P}(\Omega$ is such that $1 < p_- \leq p_+ < \infty$; suppose that for each $k \in \mathbb{N}$, $q_k \in \mathcal{P}(\Omega)$ and $\varepsilon_k > 0$ are such that $\lim_{k\to\infty} \varepsilon_k = 0$ and*

$$p(x) \leq q_k(x) \leq p(x) + \varepsilon_k \quad (x \in \Omega, k \in \mathbb{N}).$$

Let id_k be the natural embedding of $L_{q_k(\cdot)}(\Omega)$ in $L_p(\Omega)$. Then

$$\lim_{k\to\infty} \|id_k\| = 1.$$

Spaces with variable exponent also resemble classical Lebesgue spaces with regard to Lebesgue points. We recall that if p is a constant and $1 \le p < \infty$, then given any $f \in L_p(\mathbb{R}^n)$, a point $x \in \mathbb{R}^n$ such that

$$\lim_{r \to 0} |B(x,r)|^{-1} \int_{B(x,r)} |f(y) - f(x)|^p \, dy = 0$$

is called a Lebesgue point of f. It is a very familiar fact (see [37], p.44, Corollary 1) that almost every point of $\mathbb{R}^n$ is a Lebesgue point of f. By a simple adaptation of the Corollary just mentioned it can be shown that if $p \in \mathcal{P}(\mathbb{R}^n)$, $p_+ < \infty$ and $f \in L_p(\mathbb{R}^n)$, then

$$\lim_{r \to 0} |B(x,r)|^{-1} \int_{B(x,r)} |f(y) - f(x)|^{p(y)} \, dy = 0$$

for a.e. $x \in \mathbb{R}^n$.

Finally we turn to the determination of duality maps on $L_p(\Omega)$. When p is a constant in the interval $(1, \infty)$, it is a familiar fact that the Gâteaux derivative of $\|\cdot\|_p$ is given by

$$\left(\operatorname{grad} \|f\|_p \right)(x) = \|f\|_p^{1-p} |f(x)|^{p-1} \operatorname{sgn} f(x) \quad (f \ne 0, x \in \Omega),$$

from which the duality map J with gauge function μ may be calculated by means of the formula

$$Jf = \mu \left(\|f\|_p \right) \operatorname{grad} \|f\|_p .$$

Now suppose that $p \in \mathcal{P}(\Omega)$ and $1 < p_- \le p_+ < \infty$. By Theorems 1.2 and 1.8, $L_p(\Omega)$ is uniformly convex and uniformly smooth, and so its norm is even Fréchet-differentiable.

Theorem 1.14. *Let Ω be a bounded open subset of $\mathbb{R}^n$ and let $p \in \mathcal{P}(\Omega)$ be such that $1 < p_- \le p_+ < \infty$. Then for every $f \in L_p(\Omega) \backslash \{0\}$,*

$$(\operatorname{grad} \|f\|_p)(x) = \frac{p(x) \|f\|_p^{-p(x)} |f(x)|^{p(x)-1} \operatorname{sgn} f(x)}{\int_\Omega p(x) \|f\|_p^{-p(x)-1} |f(x)|^{p(x)} \, dx}. \tag{1.10}$$

Proof. Put

$$A(x) = \left(\frac{|f(x)|}{\|f\|_p} \right)^{p(x)-1} \operatorname{sgn} f(x), \quad B = \int_\Omega p(x) \|f\|_p^{-p(x)-1} |f(x)|^{p(x)} \, dx.$$

Thus the right-hand side of (1.10) equals $p(x) \|f\|_p^{-1} A(x)/B$. Note that

$$\frac{p_-}{\|f\|_p} = \frac{p_-}{\|f\|_p} \phi_p \left(f/ \|f\|_p \right) \le B \le \frac{p_+}{\|f\|_p}.$$

Moreover,

$$\phi_{p'}(A) = \int_\Omega \left(\frac{|f(x)|}{\|f\|_p} \right)^{p(x)} dx = 1.$$

Hence the right-hand side of (1.10) represents an element of $L_{p'}(\Omega)$ and so can be identified with an element of the dual of $L_p(\Omega)$, the value of which at f is $\|f\|_p$. The result follows. $\qquad\square$

This result was first established in [26].

Our discussion so far has been directed towards those aspects of the theory of spaces with variable exponent that emphasise the similarities between the properties of such spaces and those of their classical counterparts. Remarkable though these are, important differences remain, even when the exponent p is non-pathological. The most prominent of these relates to the matter of p-mean continuity. Recall that if p is a constant, $p \in [1, \infty)$, and Ω is a non-empty open bounded subset of $\mathbb{R}^n$, then every function $f \in L_p(\Omega)$, extended by 0 outside Ω, is p-mean continuous in the sense that given any $\varepsilon > 0$, there exists $\delta > 0$ such that

$$\left(\int_\Omega |f(x+h) - f(x)|^p \right)^{1/p} < \varepsilon \text{ if } |h| < \delta.$$

By analogy with this, when p is any element of $\mathcal{P}(\Omega)$, we say that a function $f \in L_p(\Omega)$ is p-mean continuous if $\rho_p(f - \tau_h f) < \varepsilon$ whenever $|h| < \delta$: here $\tau_h f := f(\cdot + h)$. It turns out that elements of $L_p(\Omega)$ are not, in general, p-mean continuous: more precisely, in [55] the following theorem is given.

Theorem 1.15. *Let $p \in \mathcal{P}(\Omega)$ and suppose that Ω contains a ball $B(x_0, r)$ on which p is continuous and non-constant. Then there is a function $f \in L_p(\Omega)$ which is not p-mean continuous.*

Proof. Let $z \in B(x_0, r)$ be a point at which p does not have a local extremum. There are sequences $(x_k)_{k \in \mathbb{N}}$, $(y_k)_{k \in \mathbb{N}}$ of points of $B(x_0, r)$, each converging to z, with $p(x_k) < p(z) < p(y_k)$ for all $k \in \mathbb{N}$. Since p is continuous in $B(x_0, r)$, for each $k \in \mathbb{N}$ there exists $r_k > 0$ such that

$$p(x) < \frac{1}{2}(p(z) + p(x_k)) < p(z) \text{ if } x \in B(x_k, r_k) \qquad (1.11)$$

and

$$p(x) > p(z) \text{ if if } x \in B(y_k, r_k). \tag{1.12}$$

For each $k \in \mathbb{N}$ put $q_k = \frac{1}{2}(p(z) + p(x_k))$, let f_k be a function on Ω such that supp $f_k \subset B(x_k, r_k)$, $f_k \in L_{q_k}(B(x_k, r_k)) \setminus L_p(B(x_k, r_k))$ and $\|f\|_{q_k} = 1$, and define

$$f = \sum_{k=1}^{\infty} 2^{-k} f_k.$$

From (1.11) and Theorem 1.12 we have

$$\|f\|_p \leq \sum_{k=1}^{\infty} 2^{-k} \|f_k\|_p \leq \sum_{k=1}^{\infty} 2^{-k} \|f_k\|_{q_k} (|B(x_k, r_k)| + 1)$$
$$\leq 1 + \sup_k |B(x_k, r_k)| < \infty.$$

Now put $h_k = y_k - x_k$ $(k \in \mathbb{N})$: from (1.12) and Theorem 1.12 we obtain

$$\|\tau_{h_k} f\|_p \geq \left\|\chi_{B(y_k, r_k)} \tau_{h_k} f\right\|_p \geq (1 + |B(y_k, r_k)|)^{-1} \left\|\chi_{B(y_k, r_k)} \tau_{h_k} f\right\|_{p(z)}$$
$$= (1 + |B(y_k, r_k)|)^{-1} \left\|\chi_{B(y_k, r_k)} f\right\|_{p(z)} = \infty.$$

It follows that $\tau_{h_k} f - f \notin L_p(\Omega)$. Thus f is not p-mean continuous. □

Despite this result, if $p_+ < \infty$ and $f \in L_p(\Omega)$ is bounded, with compact support, then there is a compact set K such that for all $h \in \mathbb{R}^n$ with $|h| < 1$, supp $\tau_h f \subset K$. Thus by Theorem 1.12,

$$\lim_{|h| \to 0} \|\tau_h f - f\|_p \leq (1 + |K|) \lim_{|h| \to 0} \|\tau_h f - f\|_{p_+} = 0.$$

The lack of satisfactory behaviour under translations is perhaps the biggest technical difficulty faced when dealing with spaces with variable exponent. Naturally it manifests itself in connection with convolutions. For example, if p is a constant, $p \in [1, \infty]$, then a special case of Young's inequality asserts that

$$\| f * g \|_{p, \mathbb{R}^n} \leq \|f\|_{p, \mathbb{R}^n} \|g\|_{1, \mathbb{R}^n}$$

for all $f \in L_p(\mathbb{R}^n)$ and $g \in L_1(\mathbb{R}^n)$; here $f * g$ denotes the convolution of f and g. However (see [23]) if $p \in \mathcal{P}(\Omega)$ and $1 < p_- \leq p_+ < \infty$, then there exists $c > 0$ such that

$$\| f * g \|_{p, \mathbb{R}^n} \leq c \|f\|_{p, \mathbb{R}^n} \|g\|_{1, \mathbb{R}^n}$$

for all $f \in L_p(\mathbb{R}^n)$ and all $g \in L_1(\mathbb{R}^n)$ if and only if p is constant. A more general form of Young's inequality for constant exponents asserts that

$$\| f * g \|_{r,\Omega} \le c \|f\|_{p,\Omega} \|g\|_{q,\Omega},$$

when $\Omega \subset \mathbb{R}^n$ and p, q, r are constants lying in the interval $[1, \infty]$ and satisfying the equality

$$\frac{1}{r} + 1 = \frac{1}{p} + \frac{1}{q}.$$

Further insight into the severe restrictions that have to be imposed in order to obtain an analogue of the classical results is provided by the following result (see [22]), which also illustrates the usefulness of Theorem 1.15.

Theorem 1.16. *Let $p, r \in \mathcal{P}(\Omega) \cap C(\Omega)$ be such that $1 < p_- \le p_+ < \infty$ and $1 < r_- \le r_+ < \infty$; suppose that Ω is bounded and contains a ball on which r is not constant. Then*

$$\| f * g \|_{r,\Omega} \lesssim \|f\|_{p,\Omega} \|g\|_{1,\Omega} \quad (f \in L_p(\Omega), g \in L_1(\Omega))$$

if and only if $p_- \ge r_+$.

Proof. First suppose that $p_- \ge r_+$. Then from the classical form of Young's inequality and Theorem 1.12,

$$\| f * g \|_{r,\Omega} \lesssim \| f * g \|_{r_+,\Omega} \lesssim \|f\|_{r_+,\Omega} \|g\|_{1,\Omega} \lesssim \|f\|_{p,\Omega} \|g\|_{1,\Omega}.$$

For the converse, assume that $\| f * g \|_{r,\Omega} \lesssim \|f\|_{p,\Omega} \|g\|_{1,\Omega}$ and suppose that $p_- < r_+$. It is sufficient to deal with the case in whcih Ω is a ball on which r is not constant and $p(x) < r(x)$. Since $L_r(\Omega) \hookrightarrow L_p(\Omega)$ (by Theorem 1.12), it follows from Theorem 1.15 that there is a function $f \in L_p(\Omega)$ which is not r-mean continuous. Hence there exists $h \in \mathbb{R}^n$ such that, in the notation used in Theorem 1.15, $f_h \notin L_r(\Omega)$. Now let ϕ be a Friedrichs mollifier and for each $\varepsilon > 0$ define ϕ_ε by $\phi_\varepsilon(x) = \varepsilon^{-n}\phi((x - h)/\varepsilon)$, $x \in \mathbb{R}^n$. Then as $\varepsilon \to 0$, $f * \phi_\varepsilon \to f_h$ in $L_1(\Omega)$; and by the continuity of the convolution operator,

$$\| f * \phi_\varepsilon \|_{r,\Omega} \lesssim \|f\|_{p,\Omega} \|\phi_\varepsilon\|_{1,\Omega} = \|f\|_{p,\Omega}.$$

As $L_r(\Omega)$ is reflexive, there is a sequence (ε_k), with $\varepsilon_k \to 0$, such that $(f * \phi_{\varepsilon_k})$ converges weakly in $L_r(\Omega)$; evidently the limit is f_h. But $f_h \notin L_r(\Omega)$ and we have a contradiction. □

Yet another difficulty arises in connection with the Hardy-Littlewood maximal operator M, defined for all functions f that are locally integrable on Ω, by

$$(Mf)(x) = \sup |B|^{-1} \int_{B \cap \Omega} |f(y)| \, dy \ (x \in \Omega),$$

where the supremum is taken over all balls B that contain x and for which $|B \cap \Omega| > 0$. It is well known that if p is a constant lying in the interval $(1, \infty)$, then M maps $L_p(\Omega)$ boundedly into itself: there is a positive constant $c = c(p, \Omega)$ such that for all $f \in L_p(\Omega)$,

$$\|Mf\|_p \leq c \|f\|_p. \tag{1.13}$$

However, (1.13) does not hold for all $p \in \mathcal{P}(\Omega)$. A sufficient condition for its validity, when Ω is open and bounded, is that for some constant $C > 0$,

$$|p(x) - p(y)| \leq -\frac{C}{\log |x - y|} \text{ for all } x, y \in \Omega \text{ with } 0 < |x - y| < 1/2.$$

A detailed discussion of this is given in [23].

Notes

The literature on spaces with variable exponent is now large and varied; no attempt is made here to give a comprehensive list of references. However, we remark that the paper of Kováčik and Rákosník [55] is of fundamental importance and that we have based several of our proofs on it; for a very detailed account of the underlying theory from a general point of view, see [23]. The elegant proof that gives equivalent conditions under which $L_p(\Omega)$ is uniformly convex is due to Lukeš, Pick and Pokorný [64]. For encyclopedic coverage of the general area we refer to the book [23]; see also [18] for detailed coverage of certain aspects of the theory.

Chapter 2

Sobolev Spaces with Variable Exponent

2.1 Definition and Functional-analytic Properties

This Chapter is intended to provide a brief summary of the main properties of Sobolev spaces with variable integrability.

Let $\Omega \subset \mathbb{R}^n$ be a bounded domain with a Lipschitz boundary and let $p \in \mathcal{P}(\Omega)$. As is customary, $\nabla u = \left(\frac{\partial}{\partial x_j} u \right)_{1 \le j \le n}$ stands for the (weak) gradient of $u \in L_p(\Omega)$. We also set $p_- = \inf_\Omega p \le \sup_\Omega p = p_+$.

Definition 2.1. The Sobolev space $W_p^1(\Omega)$ associated to the variable exponent p is defined as

$$W_p^1(\Omega) = \left\{ u \in L_p(\Omega) : |\nabla u| \in L_p(\Omega) \right\},$$

endowed with the norm

$$\|u\|_{W_p^1(\Omega)} = \|u\|_{1,p} = \|u\|_{p,\Omega} + \||\nabla u|\|_{p,\Omega}.$$

Likewise, in analogy with the standard terminology used when p is constant, $\overset{\circ}{W}_p^1(\Omega)$ will, in the sequel, stand for the closure of $C_0^\infty(\Omega)$ in $W_p^1(\Omega)$.

If there is little chance of confusion we shall use $W_p^1(\Omega)$ or W_p^1 and $\|u\|_{W_p^1(\Omega)}$ or $\|u\|_{1,p,\Omega}$.

It is apparent from the above definition that these spaces coincide with the classical Sobolev spaces if p is constant a.e in Ω. It is not surprising, on the other hand, that the validity of well-known, standard properties of the classical spaces defined for constant p will, in this general setting of variable p, depend on the quality of the function p. In this regard we highlight the following Theorem (see [55]):

Theorem 2.1. *If p is bounded a.e. in Ω, the Sobolev space $W_p^1(\Omega)$ is a separable Banach space. If, in addition, $p_- > 1$, then $W_p^1(\Omega)$ is reflexive.*

41

When furnished with the norm

$$\|u\|_{1,p} = \left(\|u\|_p^p + \sum_{k=1}^n \left\| \frac{\partial u}{\partial x_k} \right\|_p^p \right)^{\frac{1}{p}}, \tag{2.1}$$

the Sobolev space $W_p^1(\Omega)$ is, in addition, uniformly convex.

Proof. By Theorem 1.7, given a Cauchy sequence $(u_k)_{k\in\mathbb{N}}$ in $W_p^1(\Omega)$, for each subindex $j = 1, 2, ...n$, the sequence $(u_k)_{k\in\mathbb{N}}$ and the sequence of partial derivatives $(\partial u_k/\partial x_j)_{k\in\mathbb{N}}$ converge in $L_p(\Omega)$ to $u \in L_p(\Omega)$ and $v_j \in L_p(\Omega)$, respectively. A routine application of Lebesgue's dominated convergence theorem shows that $u \in W_p^1(\Omega)$ and $\frac{\partial}{\partial x_j} u = v_j$. Thus $W_p^1(\Omega)$ is a Banach space. On the other hand, it is clear by the action of the map

$$F : W_p^1(\Omega) \to L_p(\Omega) \times (L_p(\Omega))^n,$$

that $W_p^1(\Omega)$ endowed with the norm (2.1) is isometrically isomorphic to a closed subspace of a separable, uniformly convex space, provided that $1 < p_- \le p_+ < \infty$, i.e., it is separable and uniformly convex. In particular, $W_p^1(\Omega)$ is reflexive. $\qquad\square$

2.2 Sobolev Embeddings

Embedding-type theorems are central in the study of spaces of differentiable functions, due in part to their implications for the analysis of partial differential equations. The classical Sobolev embedding theorems have their counterparts in the setting of variable integrability. This section draws on results in [23] and [55]. Throughout this section Ω will be assumed to be a bounded, Lipschitz domain in $\mathbb{R}^n$.

As an immediate corollary of Theorem 1.12 we have

Theorem 2.2. *Let $|\Omega| < \infty$ and suppose that $p, q \in \mathcal{P}(\Omega)$ are such that*

$$p(x) \le q(x)$$

a.e. in Ω. Then $W_q^1(\Omega)$ is continuously embedded in $W_p^1(\Omega)$, and the norm of the embedding map id satisfies

$$\|id\| \le 1 + |\Omega|.$$

Likewise, one has the bounded embedding

$$\overset{\circ}{W}_q^1(\Omega) \hookrightarrow \overset{\circ}{W}_p^1(\Omega)$$

and its norm does not exceed $1 + |\Omega|$.

The next Sobolev embedding theorems for continuous variable exponent p are based on Theorem 2.2.

Theorem 2.3. *Let $\Omega \subset \mathbb{R}^n$ be a bounded Lipschitz domain, $n > 1$ and $p \in C(\overline{\Omega})$. If $1 < p_- \le p_+ < n$ and $0 < \varepsilon < 1/(n-1)$, there exists a positive constant C depending only on Ω and p such that for all $u \in \overset{\circ}{W}{}^1_p(\Omega)$,*

$$\|u\|_{\frac{np}{n-p}-\varepsilon} \le C\|u\|_{1,p}.$$

In other words, the embedding

$$\overset{\circ}{W}{}^1_p(\Omega) \hookrightarrow L_{\frac{np}{n-p}-\varepsilon}(\Omega) \tag{2.2}$$

is bounded.

Proof. The function

$$S : [1,n) \to [n/(n-1),\infty), \tag{2.3}$$

$$S(t) = \frac{nt}{n-t}$$

is continuous and strictly increasing. Pick $\delta > 0$ such that $p_- - \delta > 1$ and $p_+ + \delta < n$. Let $p_1 = p_- - \delta$, $r_1 = S^{-1}(S(1) + \varepsilon)$, $p_2 \in (p_1, r_1)$ and $r_2 = S^{-1}(S(p_2) + \varepsilon)$; for $j \ge 3$ set

$$p_j \in (r_{j-2}, r_{j-1}) \text{ and } r_j = S^{-1}(S(p_j) + \varepsilon).$$

Evidently, $r_j \to n$ as $j \to \infty$; let r_{N+1} be the first term of the sequence (r_j) which exceeds p_+ and set $r_N = p_+ + \delta$. Because of the continuity of p it is immediate that the family

$$p^{-1}((p_1, r_1)), p^{-1}((p_2, r_2)), ..., p^{-1}((p_j, r_j)),$$
$$..., p^{-1}((p_{N-1}, r_{N-1})), p^{-1}((p_N, p_+ + \delta))$$

is an open covering of Ω; let (φ_j) be a partition of unity associated to it. For any $u \in \overset{\circ}{W}{}^1_p(\Omega)$, Hölder's inequality (Theorem 1.5) shows that there exists a positive constant C, depending only on p, for which

$$\int_\Omega |u\varphi_j|^{p_j} \le C \||u\varphi_j|^{p_j}\|_{p/p_j} \|1\|_{p/(p-p_j)}. \tag{2.4}$$

On the other hand, by definition of the Luxemburg norm, we have

$$\int_\Omega \left(\frac{|u\varphi_j|^{p_j}}{\||u\varphi_j|\|_p^{p_j}}\right)^{p/p_j} = \int_\Omega \frac{|u\varphi_j|^p}{\|u\varphi_j\|_p^p} = 1$$

and hence, by the same token,

$$\| |u\varphi_j|^{p_j} \|_{p/p_j} \le \| u\varphi_j \|_p^{p_j}.$$

In conjunction with (2.4) the latter inequality implies the following estimate:

$$\int_\Omega |u\varphi_j|^{p_j} \le C \| u\varphi_j \|_p^{p_j} |\Omega|^\alpha, \tag{2.5}$$

where $\alpha > 0$ depends only on p. From the convexity of the modular we obtain

$$\int_\Omega |\nabla u\varphi_j|^{p_j} \le 2^{p_+ - 1} \left(\int_\Omega |\varphi_j \nabla u|^{p_j} + \int_\Omega |u\nabla\varphi_j|^{p_j} \right).$$

After estimating each of the last two integrals in the same way as above it is easy to see that there exists a positive constant C, independent of v, for which

$$\int_\Omega |\nabla u\varphi_j|^{p_j} \le C \left(\|u\|_p^{p_+} + \|\nabla u\|_p^{p_+} \right). \tag{2.6}$$

We conclude that for arbitrary $u \in \overset{\circ}{W}{}^1_{p_j}(\Omega)$ and for each $j = 1, 2, ...N$, one has

$$u\varphi_j \in \overset{\circ}{W}{}^1_{p_j}(\Omega).$$

It is inherent to the construction of the sequence (r_j) that

$$\frac{nr_j}{n - r_j} - \frac{np(x)}{n - p(x)} < \varepsilon,$$

whence the continuity of the embedding

$$\overset{\circ}{W}{}^1_{p_j}(\Omega) \hookrightarrow L_{\frac{np_j}{n-p_j}}(\Omega) = L_{\frac{nr_j}{n-r_j} - \varepsilon}(\Omega)$$

guarantees the existence of a positive constant C such that for all $u \in \overset{\circ}{W}{}^1_p(\Omega)$ and all $j = 1, 2, ...N$ the inequality

$$\| u\varphi_j \|_{nr_j/(n-r_j) - \varepsilon} \le C \|u\|_{1,p_j}$$

holds.

Next, for $u \in \overset{\circ}{W}{}^1_p(\Omega)$ it is immediate that

$$\| u \|_{\frac{np}{n-p} - \varepsilon} \le \sum_0^N \| u\varphi_j \|_{\frac{np}{n-p} - \varepsilon} \le C \sum_0^N \| u\varphi_j \|_{1,p_j} \le C \sum_0^N \| u\varphi_j \|_{1,p}$$

$$\le C \|u\|_{1,p},$$

which yields the boundedness of the embedding (2.2). $\square$

It is important to note that the continuity of p cannot be relaxed in Theorem 2.3. To demonstrate this point we present the following modification of Example 3.2 in [55].

Let Ω be the open unit disc in $\mathbb{R}^2$, let $1 < r < s < 2$; as is customary denote the polar coordinates of the point (x, y) by (ϱ, θ) and set

$$\Omega_1 = \{(\varrho, \theta) : 0 < \theta < \varrho^{2(s-r)/r}\}.$$

Choose a positive number ε such that

$$s^2 - (s\varepsilon + 1)r > 0 \ , \ \ s - r(\varepsilon + 1) > 0 \ , \ \ s^2 - rs(\varepsilon + 1) > 0.$$

Set $\gamma = \frac{s-2}{r} - \varepsilon$. Consider the function $u : \Omega \to (0, \infty)$ defined by

$$u(x) = |x|^{2(s-r)/r}.$$

If the exponent p is given by

$$p(x) = r\chi_{\Omega\setminus\Omega_1}(x) + s\chi_{\Omega_1}(x),$$

the requirements on ε are easily seen to imply that $\varphi u \in \overset{\circ}{W}{}^1_p(\Omega)$ for any cut-off function $\varphi \in C_0^\infty(\Omega)$ identically equal to 1 on $B(0, \frac{1}{2})$. In the interest of self-containedness we present the details of the proof of this fact. It suffices to show that $u \in W_p^1(\Omega)$. To that effect, we observe that

$$\rho_p(u) = \int_\Omega |x|^{p(\frac{s-2}{r}-\varepsilon)} = \int_{\Omega_1} |x|^{s(\frac{s-2}{r}-\varepsilon)} + \int_{\Omega\setminus\Omega_1} |x|^{r(\frac{s-2}{r}-\varepsilon)}$$

$$= \int_0^1 \varrho^{1+s(\frac{s-2}{r}-\varepsilon)} \int_0^{\varrho^{2(s-r)/r}} d\theta \, d\varrho + \int_0^1 \varrho^{1+r(\frac{s-2}{r}-\varepsilon)} \int_{\varrho^{2(s-r)/r}}^\pi d\theta \, d\varrho$$

$$\leq \int_0^1 \varrho^{\frac{s^2-r(\varepsilon s+1)}{r}} d\varrho + \pi \int_0^1 \varrho^{s-1-r\varepsilon} d\varrho < \infty.$$

Likewise,

$$\rho_p(|\nabla u|) = \int_\Omega |x|^{p(\frac{s-2}{r}-\varepsilon-1)} = \int_{\Omega_1} |x|^{s(\frac{s-2}{r}-\varepsilon-1)} + \int_{\Omega\setminus\Omega_1} |x|^{r(\frac{s-2}{r}-\varepsilon-1)}$$

$$= \int_0^1 \varrho^{1+s(\frac{s-2}{r}-\varepsilon-1)} \int_0^{\varrho^{2(s-r)/r}} d\theta \, d\varrho \hspace{3cm} (2.7)$$

$$+ \int_0^1 \varrho^{1+r(\frac{s-2}{r}-\varepsilon-1)} \int_{\varrho^{2(s-r)/r}}^\pi d\theta \, d\varrho$$

$$\leq \int_0^1 \varrho^{\frac{s^2-rs(\varepsilon+1)-r}{r}} d\varrho + \pi \int_0^1 \varrho^{s-1-r(\varepsilon+1)} d\varrho < \infty,$$

which proves our assertion. However, the Sobolev conjugate of p is

$$q = \frac{2r}{2-r}\chi_{\Omega\backslash\Omega_1} + \frac{2s}{s-2}\chi_{\Omega_1};$$

accordingly

$$\rho_q(u) = \int_\Omega |x|^{q(\frac{s-2}{r}-\varepsilon)} = \int_{\Omega_1} |x|^{(2s/(2-s))(\frac{s-2}{r}-\varepsilon)} + \int_{\Omega\backslash\Omega_1} |x|^{(2r/(2-r))(\frac{s-2}{r}-\varepsilon)}.$$

The first term in the right-hand side above is, however

$$\int_0^1 \varrho^{-(1+2s\varepsilon/(2-s))} = \infty.$$

Thus,

$$\overset{\circ}{W}{}^1_p(\Omega) \not\subseteq L^{np/(n-p)-\varepsilon}(\Omega).$$

2.3   Compact Embeddings

Because of their fundamental role in the analysis of existence and regularity of partial differential equations and their ubiquity in linear and non-linear analysis, as well as by reason of their theoretical importance, compact Sobolev embeddings deserve special attention. Fortunately most of the classical theorems survive in useful form in the context of variable exponent spaces. Throughout this section Ω is again assumed to be a bounded Lipschitz domain in $\mathbb{R}^n$.

Theorem 2.4. *If $p \in C(\overline{\Omega})$ satisfies $1 < p_- \leq p_+ < n-\delta$ for some positive number δ, then the embedding*

$$\overset{\circ}{W}{}^1_p(\Omega) \hookrightarrow L_{np/(n-p)-\delta}(\Omega)$$

is compact.

Proof.   We refer to the notation in the proof of Theorem 2.3. In particular, let $\epsilon < \min\{\frac{1}{n-1}, \delta\}$. Let (v_k) be a bounded sequence in $\overset{\circ}{W}{}^1_p(\Omega)$. Proposition 1.5 in conjunction with inequalities (2.5) and (2.6) immediately yields that for each $j = 1, ..., N$, the sequence $(v_k\varphi_j)_k$ is bounded in $\overset{\circ}{W}{}^1_{p_j}(\Omega)$; hence as a direct consequence of the classical Sobolev embedding theorem for constant exponent, $(v_k\varphi_j)_k$ is a Cauchy sequence in $L_{S(p_j)}(\Omega) = L_{S(r_j)-\varepsilon}(\Omega)$.

Thus, for some positive constant C independent of (v_k),

$$\|v_k - v_i\|_{np/(n-p)-\varepsilon} \leq \sum_1^N \|(v_k - v_i)\varphi_j\|_{np/(n-p)-\varepsilon}$$

$$\leq C \sum_1^N \|(v_k - v_i)\varphi_j\|_{nr_j/(n-r_j)-\varepsilon}$$

$$\leq C \sum_1^N \|(v_k - v_i)\varphi_j\|_{np_j/(n-p_j)},$$

which shows that the original sequence (v_k) is in fact a Cauchy sequence in $L_{np/(n-p)-\varepsilon}(\Omega)$. This assertion completes the proof of the Theorem. $\quad\square$

We now establish what is perhaps the most important compactness result in this section.

Theorem 2.5. *Let $p \in C(\overline{\Omega})$; assume that $n > 1$ and $1 < p_- \leq p_+ < \infty$. Then the embedding*

$$\overset{\circ}{W}{}^1_p(\Omega) \hookrightarrow L_p(\Omega)$$

is compact.

Corollary 2.1. *(Poincaré's inequality) Under the assumptions of Theorem 2.5, there exists a constant $C > 0$, independent of u, such that the inequality*

$$\|u\|_p \leq C \|\nabla u\|_p$$

holds for any $u \in \overset{\circ}{W}{}^1_p(\Omega)$.

Proof. To prove Theorem 2.5, assume first that $p_- < n$. Choose $0 < \varepsilon < \min\{\frac{1}{n-1}, \frac{p_-^2}{n-p_-}\}$. This implies that for $p(x) < n$ the condition

$$\varepsilon < \frac{p^2(x)}{n - p(x)}$$

holds, which in turn implies

$$p(x) < \frac{np(x)}{n - p(x)} - \varepsilon.$$

Let S stand for the function defined in (2.3). Select a positive number δ such that $n + \delta < \frac{n(n-\delta)}{\delta}$, $p_- - \delta > 1$ and that

$$n + \delta < \frac{n(n + \delta)}{\delta}.$$

Let $p_1 = p_- - \delta$, $r_1 = S^{-1}(S(p_1) + \varepsilon)$, $p_2 \in (p_1, r_1)$ and $r_2 = S^{-1}(S(p_2) + \varepsilon)$; for $j \geq 3$ set

$$p_j \in (r_{j-2}, r_{j-1}) \text{ and } r_j = S^{-1}(S(p_j) + \varepsilon).$$

Evidently,

$$r_j \longrightarrow n \text{ as } j \longrightarrow \infty;$$

let r_M be the first term of the sequence (r_j) that exceeds $n - \delta$. As in the proof of Theorem 2.3 the continuity of p implies that the collection

$$p^{-1}((p_1, r_1)), p^{-1}((p_2, r_2)), ..., p^{-1}((p_j, r_j)),$$

$$..., p^{-1}((p_M, r_M)), p^{-1}((n - \delta, n + \delta)), p^{-1}\left(n + \frac{\delta}{2}, \infty\right)$$

is an open covering of Ω. For the sake of simplicity we will write

$$p_{M+1} = n - \delta \ , \ r_{M+1} = n + \delta \ , \ p_{M+2} = n + \frac{\delta}{2} \ , \ r_{M+2} = \infty$$

and

$$\Omega_k = p^{-1}((p_k, r_k)).$$

Let (φ_j) be a partition of unity associated to the open covering $(\Omega_k)_{1 \leq k \leq M+2}$. In the spirit of the proof of Theorem 2.3, we take a sequence (f_i) in the unit ball of $\overset{\circ}{W}{}_p^1(\Omega)$. For fixed j with $1 \leq j \leq M$, we set

$$p_j \leq \chi_{p^{-1}((p_j, r_j)))} p + \chi_{\Omega \setminus p^{-1}((p_j, r_j))} p_+ = q;$$

in the light of Theorem 2.2, the inclusion

$$\overset{\circ}{W}{}_q^1(\Omega) \hookrightarrow \overset{\circ}{W}{}_{p_j}^1(\Omega)$$

is continuous. On the other hand, we see, by definition of the Luxemburg norm, that for any $f \in \overset{\circ}{W}{}_p^1(\Omega)$,

$$\|f\varphi_j\|_{1,q,\Omega} \leq \|f\varphi_j\|_{1,p,\Omega}.$$

Thus

$$(f_i\varphi_j)_i \subset \overset{\circ}{W}{}_{p_j}^1(\Omega);$$

and, uniformly in i, we have

$$\|f_i\varphi_j\|_{1,p_j,\Omega} \leq C.$$

Since

$$\overset{\circ}{W}{}^1_{p_j}(\Omega) \hookrightarrow\hookrightarrow L_{\frac{np_j}{n-p_j}}(\Omega)$$

one can extract a subsequence of $(f_i \varphi_j)$ (as is customary still denoted by $(f_i \varphi_j)_i$) that converges in $L_{\frac{np_j}{n-p_j}}(\Omega) = L_{\frac{nr_j}{n-r_j}-\varepsilon}(\Omega)$. The choice of p_j, r_j and the continuity of the inclusions

$$L_{\frac{nr_j}{n-r_j}-\varepsilon}(\Omega) \hookrightarrow L_{\frac{np}{n-p}-\varepsilon}(\Omega) \hookrightarrow L_p(\Omega)$$

yield the following string of inequalities (in which C denotes a positive constant independent of the sequence (f_i) and is not necessarily the same in each occurrence) for the subsequence under consideration:

$$\int_\Omega |f_i\varphi_j - f_k\varphi_j|^p = \int_{p^{-1}((p_j,r_j))} |f_i\varphi_j - f_k\varphi_j|^p$$

$$\leq C \int_{p^{-1}((p_j,r_j))} |f_i\varphi_j - f_k\varphi_j|^{\frac{np}{n-p}-\varepsilon}$$

$$\leq C \int_{p^{-1}((p_j,r_j))} |f_i\varphi_j - f_k\varphi_j|^{\frac{nr_j}{n-r_j}-\varepsilon}$$

$$= C \int_{p^{-1}((p_j,r_j))} |f_i\varphi_j - f_k\varphi_j|^{\frac{np_j}{n-p_j}}$$

$$\leq C \int_\Omega |f_i\varphi_j - f_k\varphi_j|^{\frac{np_j}{n-p_j}} \leq C \|f_i\varphi_j - f_k\varphi_j\|^{p_j}_{1,p_j,\Omega}$$

$$\leq C \|f_i\varphi_j - f_k\varphi_j\|^{p_j}_{1,p,\Omega}.$$

In view of Proposition 1.5 it is immediate then that $(f_i\varphi_j)_i$ converges in $L_p(\Omega)$. Choose a subsequence of (f_i) such that $(f_i\varphi_1)$ converges in $L_{\frac{n}{n-p}-\varepsilon}(\Omega)$; the preceding reasoning shows that this subsequence still has a subsequence such that $(f_i\varphi_2)$ converges in $L_{\frac{np}{n-p}-\varepsilon}(\Omega)$; continuing with this process for $j = 1, 2 \dots M$ it is easy to construct a subsequence (f_i) of the original sequence such that for each $j = 1, \dots, M$,

$$(f_i\varphi_j)_i \tag{2.8}$$

is Cauchy in $L_{\frac{np}{n-p}-\varepsilon}(\Omega)$. In what follows we refer to this sequence. Since φ_{M+1} is supported in $p^{-1}((n-\delta, n+\delta))$, setting

$$q = (n-\delta)\chi_{p^{-1}((n-\delta,n+\delta))} + p_+ \chi_{\Omega \setminus p^{-1}((n-\delta,n+\delta))}$$

and invoking again Theorem 2.2 it is easy to see that, for some positive constant C independent of f, the inequality

$$\|f\varphi_{M+1}\|_{1,n-\delta,\Omega} \leq \|f\varphi_{M+1}\|_{1,q,\Omega} \leq C\|f\varphi_{M+1}\|_{1,p,\Omega}$$

holds for all $f \in L_p(\Omega)$. In particular, if (f_i) is the sequence alluded to in (2.8) it can be assumed that $(f_i \varphi_{M+1})_i$ is bounded in $\overset{\circ}{W}^1_{n-\delta}(\Omega)$, whence by virtue of the classical Sobolev embedding theorem

$$\overset{\circ}{W}^1_{n-\delta}(\Omega) \hookrightarrow\hookrightarrow L_{n(n-\delta)/\delta}(\Omega)$$

the sequence $(f_i \varphi_{M+1})_i$ can be assumed to be Cauchy in $L_{n(n-\delta)/\delta}(\Omega)$. Define

$$r =: n(n-\delta)/\delta \chi_{p^{-1}(n-\delta,n+\delta)} + p_+ \chi_{\Omega \setminus p^{-1}(n-\delta,n+\delta)}.$$

Then for the sequence (f_i) that satisfies (2.8) we have

$$\int_\Omega \left(\frac{|(f_i - f_k)\varphi_{M+1}|}{\|(f_i - f_k)\varphi_{M+1}\|_{n(n-\delta)/\delta}} \right)^r \tag{2.9}$$
$$= \int_{p^{-1}(n-\delta,n+\delta)} \left(\frac{|(f_i - f_k)\varphi_{M+1}|}{\|(f_i - f_k)\varphi_{M+1}\|_{n(n-\delta)/\delta}} \right)^{n(n-\delta)/\delta}$$
$$\leq \int_\Omega \left(\frac{|(f_i - f_k)\varphi_{M+1}|}{\|(f_i - f_k)\varphi_{M+1}\|_{n(n-\delta)/\delta}} \right)^{n(n-\delta)/\delta}$$
$$= 1;$$

by definition, then

$$\|(f_i - f_k)\varphi_{M+1}\|_r \leq \|(f_i - f_k)\varphi_{M+1}\|_{n(n-\delta)/\delta}.$$

In conclusion, $(f_i \varphi_{M+1})$ can be assumed to be Cauchy in $L_r(\Omega)$. Furthermore, since $p \leq r$ in Ω, Theorem 1.12 yields

$$L_r(\Omega) \hookrightarrow L_p(\Omega),$$

so that in fact $(f_i \varphi_{M+1})$ is Cauchy in $L_p(\Omega)$.

Finally, we observe that φ_{M+2} is supported in the open set $p^{-1}\left(\left(n + \frac{\delta}{2}, p_+ \right) \right)$. Set

$$s = (n + (\delta/2))\chi_{p^{-1}((n+\delta/2,\infty))} + p_- \chi_{\Omega \setminus p^{-1}((n+\delta/2,\infty))}.$$

Clearly

$$s(x) \leq p(x)$$

in Ω; Theorem 2.2 yields the continuous inclusion

$$\overset{\circ}{W}^1_p(\Omega) \hookrightarrow \overset{\circ}{W}^1_s(\Omega).$$

It can be easily verified from the definition of the Luxemburg norm that for any $f \in \overset{\circ}{W}{}^1_p(\Omega)$ the inequality

$$\|f\varphi_{M+2}\|_{1,n+\frac{\delta}{2}} \le \|f\varphi_{M+2}\|_{1,s}$$

holds. In all, the sequence $(f_i\varphi_{M+2})_i$ is bounded in $\overset{\circ}{W}{}^1_{n+(\delta/2)}(\Omega)$. The inclusion

$$\overset{\circ}{W}{}^1_{n+\delta}(\Omega) \hookrightarrow\hookrightarrow C(\overline{\Omega})$$

is, on the other hand, compact as is guaranteed by the corresponding constant-exponent Sobolev embedding theorem. Hence, without loss of generality $(f_i\varphi_{M+2})$ can also be assumed to converge in $C(\overline{\Omega})$. The boundedness of the inclusion

$$C(\overline{\Omega}) \hookrightarrow L_p(\Omega)$$

implies that $(f_i\varphi_{M+2})$ is Cauchy in $L_p(\Omega)$.

We have thus found a convergent subsequence (f_i) of the original sequence, since

$$\|f_k - f_l\|_p \le \sum_0^{M+2} \|(f_k - f_l)\varphi_j\|_p,$$

which clearly tends to 0 as k, l tend to infinity.

For the remaining case $p_- \ge n$, one clearly has:

$$\overset{\circ}{W}{}^1_p(\Omega) \hookrightarrow \overset{\circ}{W}{}^1_n(\Omega) \hookrightarrow\hookrightarrow L_p(\Omega),$$

the compactness of the last inclusion being well known.

Turning to the proof of Corollary 2.1, if no such positive constant C existed it would be possible to choose a sequence $(u_i) \in \overset{\circ}{W}{}^1_p(\Omega)$ with

$$\|\nabla u_i\|_p = 1 \quad \text{and} \quad \|u_i\|_p \longrightarrow \infty \quad \text{as } i \longrightarrow \infty;$$

set

$$v_i = u_i/\|u_i\|_{1,p}.$$

Because of the compactness of the embedding

$$\overset{\circ}{W}{}^1_p(\Omega) \hookrightarrow L_p(\Omega),$$

there exists $v \in \overset{\circ}{W}{}^1_p(\Omega)$ such that

$$v_i \rightharpoonup v \text{ in } \overset{\circ}{W}{}^1_p(\Omega) \quad \text{and} \quad v_i \to v \text{ in } L_p(\Omega).$$

Moreover, due to the normalization, one also has $\|v\|_p = 1$.

However, given any test function $\phi \in C_0^\infty(\Omega)$, for the linear functional Λ on $\overset{\circ}{W}{}_p^1(\Omega)$ defined by

$$\langle u, \Lambda \rangle = \int_\Omega u \frac{\partial \phi}{\partial x_k},$$

it is clear that

$$|\langle v_i, \Lambda \rangle| = \left| \int_\Omega \frac{\partial v_i}{\partial x_k} \phi \right| \le \|\nabla v_i\|_p \|\phi\|_{p/p-1} \longrightarrow 0 \text{ as } i \longrightarrow \infty.$$

This contradicts the fact that $v_i \rightharpoonup v \neq 0$ in $\overset{\circ}{W}{}_p^1(\Omega)$.   □

2.3.1   *Potential estimates and the class* $\mathcal{P}^{\log}(\Omega)$

A fruitful alternate approach to Sobolev embeddings and Poincaré-type inequalities relies on potential theory. Potential-theoretic methods are not only mathematically important in their own right but also constitute a connection with the realm of Calderón-Zygmund harmonic analysis, through which the scale of variable-exponent Sobolev spaces becomes amenable to the application of Calderón-Zygmund techniques. However, the key starting point in this scheme is the boundedness of the Hardy-Littlewood maximal operator on the scale $L_p(\Omega)$, which will not hold unless further restrictions are imposed on the variable exponent p, as mentioned in the paragraph following Theorem 1.15. We highlight the work [23]   for details on this vast topic. The reader is also referred to the book by Cruz-Uribe and Fiorenza [18] for an extensive account of harmonic analysis on variable-exponent Lebesgue spaces.

In what follows the admissible exponents p will be restricted to the class of globally Hölder-continuous functions, $\mathcal{P}^{\log}(\Omega)$, which we now define:

Definition 2.2. A function $v \in \mathcal{P}(\Omega)$ is said to be globally Hölder-continuous, written $v \in \mathcal{P}^{\log}(\Omega)$, if there exist $v_\infty \in \mathbb{R}$ and positive constants α, β such that for any $x \in \Omega$, $y \in \Omega$,

$$|v(x) - v(y)| \le \frac{\alpha}{\log\left(e + |x - y|^{-1}\right)}$$

and

$$|v(x) - v_\infty| \le \frac{\beta}{\log\left(e + |x|\right)}.$$

Remark 2.1. Obviously, the second condition is superfluous if Ω is bounded. For then, it will hold for some β if $v_\infty = 0$. In this case the smallest constant α for which Definition 2.2 holds is referred to as the logarithmic constant of v.

Theorem 2.6. *If* $\Omega \subset \mathbb{R}^n$ *is a bounded Lipschitz domain and* $p \in \mathcal{P}^{\log}(\Omega)$, *there exist an extension* $\tilde{p} \in \mathcal{P}^{\log}(\mathbb{R}^n)$ *and a bounded extension operator*

$$\mathcal{E} : W_p^1(\Omega) \to W_{\tilde{p}}^1(\mathbb{R}^n);$$

that is, for each $u \in W_p^1(\Omega)$, $\mathcal{E}(u)|_\Omega \equiv u$ *and*

$$\|\mathcal{E}(u)\|_{1,\tilde{p},\mathbb{R}^n} \leq C(p, \Omega, n) \|u\|_{1,p,\Omega}.$$

Moreover, $\tilde{p}_- = p_-$, $\tilde{p}_+ = p_+$ *and the logarithmic constant of* $\tilde{p}$ *is the same as that of* p.

Proof. The extension of p can be accomplished via McShane's idea ([69]), which we now sketch, for the sake of completeness. It is a routine matter to verify the concavity of the modulus of continuity

$$\omega(t) = \alpha \left(\log \left(e + t^{-1} \right) \right)^{-1};$$

in particular it follows that for all positive t and s,

$$\omega(t + s) \leq \omega(t) + \omega(s). \tag{2.10}$$

Let α be the logarithmic constant (see Remark 2.1) of p. For $x \in \mathbb{R}^n$ set $\tilde{p}(x) = p(x)$ if $x \in \Omega$ and otherwise

$$\tilde{p}(x) = \sup_{y \in \Omega} \left(p(y) - \frac{\alpha}{\log \left(e + |x - y|^{-1} \right)} \right).$$

Let

$$\overline{p}(x) = \max\{\tilde{p}(x), p_-\}.$$

For fixed $x \in \mathbb{R}^n$ and $y \in \Omega$ choose $z \in \Omega$ with $|y - z| < |x - y|$. Clearly

$$p(y) \leq |p(y) - p(z)| + p(z) \leq \frac{\alpha}{\log \left(e + |z - y|^{-1} \right)} + p_+ \leq \frac{\alpha}{\log \left(e + |x - y|^{-1} \right)} + p_+,$$

so that $\tilde{p}(x) < \infty$. For arbitrary $x, z \in \mathbb{R}^n$ and $y \in \Omega$ we have, according to (2.10),

$$\omega(|y - z|) \leq \omega(|y - x|) + \omega(|x - z|).$$

Thus,

$$p(y) - \omega(|x - y|) \leq p(y) - \omega(|y - z|) + \omega(|x - z|)$$
$$\leq \sup_{y \in \Omega} \left(p(y) - \omega(|y - z|) \right) + \omega(|x - z|);$$

therefore

$$|\tilde{p}(x) - \tilde{p}(y)| \leq \omega(|x - z|).$$

This shows that $\tilde{p}$ is log-Hölder continuous with constant α: hence if

$$p_- \leq \min\{\tilde{p}(y), \tilde{p}(x)\}$$

then

$$|\overline{p}(x) - \overline{p}(y)| \leq \frac{\alpha}{\log(e + |x - y|^{-1})}.$$

If $\tilde{p}(x) < p_- \leq \tilde{p}(y)$, however

$$|\overline{p}(x) - \overline{p}(y)| = \tilde{p}(y) - p_- \leq \tilde{p}(y) - \tilde{p}(x) \leq \omega(|y - x|).$$

It can be seen straight from the definition that $p_- \leq \overline{p} \leq p_+$; hence $\overline{p}$ satisfies the conditions on Definition 2.2 with constants α, $v_\infty = 0$ and $\beta = M p_+$, where $M = \sup_{w \in \Omega} |w|$. The extension operator $\mathcal{E}$ can be explicitly constructed via Whitney's decomposition. Since the details are somewhat lengthy, we refer the interested reader to [23] and the references therein.   □

Theorem 2.7. *If Ω is a bounded, Lipschitz domain and $p \in \mathcal{P}^{log}(\Omega)$ satisfies $1 < p_- \leq p_+ < \infty$, then both $C^\infty(\overline{\Omega})$ and $C^\infty(\Omega) \bigcap W_p^1(\Omega)$ are dense in $W_p^1(\Omega)$.*

Proof. From the boundedness of p and the inequalities in Proposition 1.5, a straightforward calculation shows that for any $u \in W_p^1(\Omega)$,

$$\lim_{k \to \infty} \|u - u_k\|_{1,p} \leq \lim_{k \to \infty} \left[\int_{\Omega \cap \{x : |u(x)| \geq k\}} (|u|^p + |\nabla u|^p) \right]^{1/p_+} = 0,$$

where $u_k = \max\{\min\{u(x), k\}, -k\} \in W_p^1(\Omega)$. Thus, bounded functions are dense in $W_p^1(\Omega)$. Furthermore, compactly supported bounded functions are dense in $W_p^1(\mathbb{R}^n)$. Indeed, given $u \in W_p^1(\mathbb{R}^n)$ and a standard cut-off function φ_r, supported in the ball $B_r(0)$ and with $0 \leq \varphi_r \leq 1$, $\varphi_r \equiv 0$ on $\mathbb{R}^n \setminus B_{2r}(0)$ and $|\nabla \varphi_r| \leq \frac{C}{r}$, we have

$$\lim_{r \to \infty} u \varphi_r = u \qquad (2.11)$$

in $W_p^1(\mathbb{R}^n)$. To prove (2.11) observe that

$$|u(1 - \varphi_r)| \leq |u| \, , \, |(1 - \varphi_r)\nabla u| \leq |\nabla u| \, , \, |\nabla u - u\nabla \varphi_r| \leq \frac{C|u|}{r} + |\nabla u|;$$

Lebesgue's dominated convergence theorem then gives

$$\lim_{r \to \infty} \int_{\mathbb{R}^n} |u(1 - \varphi_r)|^p = 0$$

and

$$\lim_{r \to \infty} \int_{\mathbb{R}^n} |\nabla u - \nabla(u\varphi_r)|^p = 0;$$

hence (2.11) follows from the boundedness of p coupled with Proposition 1.5.

Utilizing a standard mollifier argument it follows immediately from the above result that under the same assumption on the exponent p, $C_0^\infty(\mathbb{R}^n)$ is dense in $W_p^1(\mathbb{R}^n)$; indeed, setting as usual

$$\varphi(x) = c \exp\left(-1/(1 - |x|^2)\right) \chi_{B_1(0)}(x) \,, \quad \int_{\mathbb{R}^n} \varphi = 1,$$

$$\varphi_\varepsilon(x) = \frac{1}{\varepsilon^n} \varphi\left(\frac{x}{\varepsilon}\right),$$

it is well known (see [23], Section 3.6) that under the given assumptions on the exponent p, for any $u \in W_p^1(\mathbb{R}^n)$ we have

$$\varphi_\varepsilon * u \to u$$

in $W_p^1(\mathbb{R}^n)$ and that the convolution $\varphi_\varepsilon * u$ is compactly supported if so is u. It follows immediately from the density of smooth, compactly supported functions just proved that for any $u \in W_p^1(\Omega)$, the extension $\mathcal{E}(u) \in W_{\tilde{p}}^1(\mathbb{R}^n)$ given by Theorem 2.6 can be approximated by a sequence $(w_k) \subset C_0^\infty(\mathbb{R}^n)$; by definition of the Luxemburg norm it is clear that

$$\|u - w_k|_\Omega\|_{1,p,\Omega} \le \|\mathcal{E}(u) - w_k\|_{1,\tilde{p},\mathbb{R}^n}.$$

Thus, $C^\infty(\overline{\Omega})$ is dense in $W_p^1(\Omega)$. For the second part of the Theorem, fix $x_0 \in \Omega$, set $\Omega_k = \{x \in \Omega : d(x, \partial\Omega) > \frac{1}{k}\} \cap B(x_0, k)$ and set

$$U_k = \Omega_k \setminus \overline{\Omega_{k-1}};$$

let $(\chi_k)_k$ be a partition of unity subordinate to the cover $(U_k)_k$. Given an arbitrary $u \in W_p^1(\Omega)$, and a mollifier η, pick ε_k so that $\eta_{\varepsilon_k} * (\chi_k u)$ is supported in U_k. Standard arguments show that any neighbourhood of u in the topology of $W_p^1(\Omega)$ contains a function of the form

$$\sum_k \eta_{\varepsilon_k} * (\chi_k u),$$

which is in $C_0^\infty(\Omega)$. The reader is referred to [23] for further discussions of this theme. $\qquad\square$

In the course of the proof of Theorem 2.7 it has been shown that for $p \in \mathcal{P}^{\log}(\mathbb{R}^n) \bigcap L^\infty(\mathbb{R}^n)$ one has:

$$W_p^1(\mathbb{R}^n) = \overset{\circ}{W}{}_p^1(\mathbb{R}^n).$$

That the condition on the boundedness of the exponent p cannot be relaxed can be easily seen by considering $p(x) = \max\{|x|, 1\}$ and $u(x) \equiv \frac{1}{2}$ on the real line (see [23], Example 8.1.4); in fact $C_0^\infty(\mathbb{R}^n)$ is, in general, not dense in $W_p^1(\mathbb{R}^n)$, and so, in general, in contrast to the constant case,

$$W_p^1(\mathbb{R}^n) \neq \overset{\circ}{W}{}_p^1(\mathbb{R}^n).$$

2.4   Riesz Potentials

The remainder of the chapter is devoted to the analysis of Poincaré-type inequalities and the extension to variable p of the classical Sobolev embedding theorems. Much of the presentation depends on the pointwise estimate, valid a.e. in Ω for arbitrary $u \in \overset{\circ}{W}{}_p^1(\Omega)$ (to be proved in Theorem 2.10):

$$|u(x)| \leq c \int_\Omega \frac{|\nabla u(y)|}{|x - y|^{n-1}}\, dy.$$

The preceding expression motivates the introduction of the Riesz potentials. For $0 < \alpha < n$, the α-Riesz potential $I_{\Omega,\alpha} f$ of $f \in L_1(\Omega)$ is defined by

$$I_{\Omega,\alpha} f(x) = \int_\Omega \frac{f(y)}{|x - y|^{n-\alpha}}\, dy.$$

In Theorem 2.8 the boundedness of the α-Riesz potential as an operator acting on the generalised Lebesgue spaces will be established. A few technical lemmas will facilitate the exposition. As is customary, we use the symbol ω_n to represent the surface area of the unit ball in $\mathbb{R}^n$.

Lemma 2.1. *Let $p \in \mathcal{P}^{\log}(\mathbb{R}^n)$. Then there exists a positive constant C_-, depending on n and p, such that for any ball B in $\mathbb{R}^n$ and with $p_B^- = \inf_B p$, the following estimate holds for any $x \in B$:*

$$|B|^{1/p(x)-1/p_B^-} \leq C_-. \tag{2.12}$$

Proof. The left-hand side is obviously bounded by 1 if $|B| \geq 1$. If $|B| < 1$, since $|B| = \frac{\omega_n}{n} r^n < 1$, the assumption on p yields

$$|B|^{\frac{1}{p(x)} - \frac{1}{p_B}} \leq \exp\left(\frac{-C \log \frac{\omega_n}{n} r^n}{\log\left(e + (2r)^{-1}\right)}\right) = \exp\left(\frac{-C \log \frac{\omega_n}{n} - Cn \log r}{\log\left(e + (2r)^{-1}\right)}\right). \tag{2.13}$$

Clearly

$$\frac{-C \log \frac{\omega_n}{n}}{\log\left(e + (2r)^{-1}\right)} \leq \max\left\{0, \log\left(\frac{\omega_n}{n}\right)^{-C}\right\} \tag{2.14}$$

For $r < \frac{1}{3}$,

$$\frac{-Cn \log r}{\log\left(e + (2r)^{-1}\right)} \leq \frac{-Cn \log r}{\log \frac{1}{2r}} = \frac{Cn \log r}{\log 2r}, \tag{2.15}$$

whereas if $r \geq \frac{1}{3}$, the left-hand side in (2.15) is bounded by

$$\frac{Cn \log 3}{\log\left(e + 1/2r\right)} \leq Cn \log 3. \tag{2.16}$$

The estimate (2.12) for $|B| < 1$ follows immediately from (2.13), (2.14), (2.15) and (2.16). $\qquad\square$

An analogous argument shows that, under the aforementioned assumptions, there exists a positive constant C_+, depending only on the variable exponent p and the dimension n, such that the estimate

$$|B|^{1/p(x) - 1/(p_B)^+} \leq C_+ = C_+(n, p)$$

holds for any ball B, with $p_B^+ = \sup_B p$.

Lemma 2.2. *Let $p \in \mathcal{P}^{\log}(\mathbb{R}^n)$. Then there exists a positive constant C such that for any cube (or ball) B,*

$$\|\chi_B\|_p \leq C|B|^{\frac{1}{|B|} \int_B 1/p(y)\, dy}. \tag{2.17}$$

Proof. The estimate is a direct consequence of Jensen's inequality, Lemma 2.1 and the comment thereafter. To see this, we observe that if $|B| < 1$, then

$$C_- \geq |B|^{1/p(x) - 1/(p_B)^-} \geq |B|^{1/p(x)} \frac{1}{|B|} \int_B \left(\frac{1}{|B|}\right)^{1/p(y)} dy,$$

whereas if $|B| \geq 1$ the same argument gives

$$|B|^{1/p(x) - 1/(p_B)^+} \leq C_+.$$

In all,

$$|B|^{1/p(x)} \frac{1}{|B|} \int_B \left(\frac{1}{|B|} \right)^{1/p(y)} dy \leq \max\{C_-, C_+\} = M = M(n,p);$$

and, as an easy calculation shows, this implies that

$$\int_{\mathbb{R}^n} \chi_B \left(\frac{1}{M} \frac{1}{|B|} \int_B \left(\frac{1}{|B|} \right)^{1/p(y)} dy \right)^{p(x)} dx \leq 1.$$

Immediately from the definition of the L_p-norm it follows from the foregoing inequality that

$$\|\chi_B\|_p \leq M \left(\frac{1}{|B|} \int_B \left(\frac{1}{|B|} \right)^{1/p(y)} dy \right)^{-1}. \qquad (2.18)$$

On the other hand, by virtue of Jensen's inequality, for any ball B we have

$$|B|^{-\frac{1}{|B|} \int_B \frac{1}{p(y)} dy} \leq \frac{1}{|B|} \int_B \left(\frac{1}{|B|} \right)^{1/p(x)} dx,$$

and the substitution of this inequality in (2.18) yields (2.17) immediately. $\square$

Lemma 2.3. *Suppose* $p \in \mathcal{P}^{log}(\mathbb{R}^n)$ *satisfies*

$$\left| \frac{1}{p(z)} - \frac{1}{p_\infty} \right| \leq \frac{c(p)}{\log(e + |z|)} \quad (z \in \mathbb{R}^n) \qquad (2.19)$$

and $1 \leq p_- \leq p_+ < \infty$; *let* $\lambda \in (0,1)$ *satisfy the condition*

$$\frac{\log \sqrt{\lambda}}{c(p)} < -(n+1). \qquad (2.20)$$

Then the following inequality holds for any ball $B(x_0, r) \subset \mathbb{R}^n$ *and each* $x \in B(x_0, r)$:

$$\frac{1}{|B|} \int_B \lambda^{\left| \frac{1}{p(x)} - \frac{1}{p(y)} \right|^{-1}} dy \leq \frac{1}{(1+|x|)^{n+1}} + \frac{1}{|B|} \int_B \frac{1}{(1+|y|)^{n+1}} dy.$$

Proof. From the assumption on p, the inequality (2.19) is valid for all $z \in \mathbb{R}^n$ and $0 < \lambda < 1$.
Furthermore, the function

$$t \longrightarrow w(t) = t\lambda^{1/2t}$$

is easily seen to be convex. Setting

$$a(z) = 1/p(z) - 1/p_\infty$$

we have

$$(|a(x)| + |a(y)|)\, \lambda^{1/|a(x)-a(y)|} \leq 2 \left(\frac{|a(x)|}{2} + \frac{|a(y)|}{2} \right) \lambda^{1/(|a(x)|+|a(y)|)}$$

$$= 2w \left(\frac{|a(x)|}{2} + \frac{|a(y)|}{2} \right)$$

$$\leq w(|a(x)|) + w(|a(y)|)$$

$$= |a(x)| \sqrt{\lambda}^{1/|a(x)|} + |a(y)| \sqrt{\lambda}^{1/|a(y)|}.$$

Consequently

$$\lambda^{1/|a(x)-a(y)|} \leq \lambda^{1/|2a(x)|} + \lambda^{1/|2a(y)|}. \tag{2.21}$$

Furthermore, in view of condition (2.20),

$$1/|a(x)| \geq \frac{\log(1+|x|)}{c(p)} \quad \text{and} \quad 1/|a(y)| \geq \frac{\log(1+|y|)}{c(p)}.$$

The preceding inequalities in conjunction with the conditions imposed on λ imply

$$\log \left(\sqrt{\lambda} \right)^{1/|a(x)|} \leq \frac{\log \sqrt{\lambda}}{c(p)} \log (1+|x|) \leq \log (1+|x|)^{-(n+1)}$$

and

$$\log \left(\sqrt{\lambda} \right)^{1/|a(y)|} \leq \frac{\log \sqrt{\lambda}}{c(p)} \log (1+|y|) \leq \log (1+|y|)^{-(n+1)}.$$

The proof of the Lemma is completed by integrating (2.21) over the ball B and using the preceding two estimates.

$\square$

Lemma 2.3 has the following particular implication:

Lemma 2.4. *If $q \in \mathcal{P}^{log}(\mathbb{R}^n)$ and $1 \leq q_- \leq q_+ < \infty$, there exists a positive constant $C = C(q,n)$ such that if $B \subset \mathbb{R}^n$ is any ball, $x \in B$ and*

$$q_B = |B|^{-1} \int_B 1/q(z)\, dz,$$

then

$$|B|^{-\frac{q(x)}{q_B}} \leq$$

$$C \left(|B|^{-1} + (1+|x|)^{-n-1} + \left(\frac{1}{|B|} \int_B (1+|y|)^{-n-1} dy \right)^{q_-} \right).$$

Proof.   An application of Jensen's inequality to the convex function

$$\varphi(s) = |B|^{-s}$$

on the measurable space $(B, dz/|B|)$ yields

$$\frac{1}{|B|^{1/q_B}} \leq \frac{1}{|B|} \int_B \left(\frac{1}{|B|}\right)^{1/q(y)} dy.$$

Therefore

$$|B|^{-q(x)/q_B} \leq \left(\frac{1}{|B|} \int_B \left(\frac{1}{|B|}\right)^{1/q(y)} dy\right)^{q(x)}$$

$$\leq \left(\frac{1}{|B|} \int_{B \cap \{y : q(x) \geq q(y)\}} \left(\frac{1}{|B|}\right)^{1/q(y)} dy\right.$$

$$\left. + \frac{1}{|B|} \int_{B \cap \{y : q(x) < q(y)\}} \left(\frac{1}{|B|}\right)^{1/q(y)} dy\right)^{q(x)} \tag{2.22}$$

$$\leq 2^{q_+ - 1} \left(\frac{1}{|B|} \int_{B \cap \{y : q(x) \geq q(y)\}} \left(\frac{1}{|B|}\right)^{1/q(y)} dy\right)^{q(x)}$$

$$+ 2^{q_+ - 1} \left(\frac{1}{|B|} \int_{B \cap \{y : q(x) < q(y)\}} \left(\frac{1}{|B|}\right)^{1/q(y)} dy\right)^{q(x)}. \tag{2.23}$$

Using Jensen's inequality it follows readily that

$$\left(\frac{1}{|B|} \int_{B \cap \{y : q(x) \geq q(y)\}} \left(\frac{1}{|B|}\right)^{1/q(y)} dy\right)^{q(x)}$$

$$\leq \frac{1}{|B|} \int_B \left(\frac{\chi_{\{y : q(x) \geq q(y)\}}}{|B|}\right)^{q(x)/q(y)} dy. \tag{2.24}$$

If $|B| \geq 1$, it is quickly seen that the last integrand of (2.24) is bounded by $\frac{1}{|B|}$. When $|B| < 1$,

$$\left(\frac{1}{|B|}\right)^{q(x)/q(y)} \leq \left(\frac{1}{|B|}\right)^{q(x)/q_-} \leq C(q,n)\frac{1}{|B|},$$

by virtue of Lemma 2.1. In all, for arbitrary B, we have

$$\frac{1}{|B|} \int_{B \cap \{y : q(x) \geq q(y)\}} \left(\frac{1}{|B|}\right)^{q(x)/q(y)} \leq \frac{1}{|B|}.$$

If $q(x) < q(y)$, let λ be as in Lemma 2.3 and choose $w = \lambda^{q_+}$. Recall Young's inequality:

$$(ab)^s \leq \frac{s}{r}a^r + \frac{s}{t}b^t,$$

valid for $a \geq 0$, $b \geq 0$ and $1/r + 1/t = 1/s$. The substitution

$$a = \left(\frac{1}{|B|^{\alpha} w} \right)^{1/q(y)} \quad , \quad b = w^{1/q(y)}$$

and

$$\frac{1}{q(y)} + \frac{1}{q(x)} - \frac{1}{q(y)} = \frac{1}{q(x)} = \frac{1}{s}$$

gives, for any $\beta > 0$,

$$\left(\frac{1}{|B|^{\beta}} \right)^{q(x)/q(y)} = \left(\frac{w^{1/q(y)}}{(|B|^{\beta} w)^{1/q(y)}} \right)^{q(x)}$$

$$\leq C \left(\frac{1}{|B|^{\beta}} + \lambda^{1/(1/q(x)-1/q(y))} \right). \qquad (2.25)$$

Integrating (2.25), using Jensen's inequality and invoking Lemma 2.3 it follows that

$$\left(\frac{1}{|B|} \int_{B \cap \{y : q(x) < q(y)\}} \frac{1}{|B|^{\beta/q(y)}} \, dy \right)^{q(x)}$$

$$\leq \frac{1}{|B|} \int_{B \cap \{y : q(x) \leq q(y)\}} \left(\frac{1}{|B|^{\beta}} \right)^{q(x)/q(y)}$$

$$\leq C(n,q) \left(\frac{1}{|B|^{\beta}} + \frac{1}{(1+|x|)^{n+1}} + \frac{1}{|B|} \int_{B} \frac{1}{(1+|y|)^{n+1}} \, dy \right). \qquad (2.26)$$

In particular (2.26) is valid when one replaces the exponent q with q/q_- and sets $\beta = 1/q_-$. Thus

$$\left(\frac{1}{|B|} \int_{B \cap \{y : q(x) < q(y)\}} \frac{1}{|B|^{1/q(y)}} \, dy \right)^{q(x)/q_-}$$

$$\leq C \left(\frac{1}{|B|^{1/q_-}} + \frac{1}{(1+|x|)^{n+1}} + \frac{1}{|B|} \int_{B} \frac{1}{(1+|y|)^{n+1}} \, dy \right),$$

or equivalently:

$$\left(\frac{1}{|B|} \int_{B \cap \{y : q(x) < q(y)\}} \frac{1}{|B|^{1/q(y)}} \, dy \right)^{q(x)}$$

$$\leq C(q,n) \left(\frac{1}{|B|} + \frac{1}{(1+|x|)^{(n+1)q_-}} + \left(\frac{1}{|B|} \int_{B} \frac{1}{(1+|y|)^{n+1}} \, dy \right)^{q_-} \right). \qquad (2.27)$$

Lemma 2.4 follows from (2.23), (2.26) and (2.27). $\qquad \square$

We are now ready to prove the main Theorem in this section, namely the boundedness of the Riesz potentials on suitable spaces. Specifically:

Theorem 2.8. *If $p \in \mathcal{P}^{log}(\mathbb{R}^n)$, $0 < \alpha < n$ and $1 < p_- \leq p_+ < \frac{n}{\alpha}$, the linear operator*

$$I_\alpha : L_p(\mathbb{R}^n) \to L_{np/(n-\alpha p)}(\mathbb{R}^n)$$

is bounded.

Proof. Let q be the Hölder conjugate of the exponent p: for any $x \in \mathbb{R}^n$, it is obvious by Hölder's inequality that given $\delta > 0$ we have

$$\left| \int_{|x-y| \geq \delta} \frac{f(y)}{|x-y|^{n-\alpha}} \, dy \right|$$

$$\leq 2\|f\|_p \left\| \chi_{\mathbb{R}^n \setminus B(x,\delta)}(\cdot)|x - \cdot|^{\alpha-n} \right\|_q$$

$$\leq 2\|f\|_p \left\| \chi_{\mathbb{R}^n \setminus B(x,\delta)}(\cdot)|x - \cdot|^{-n} \right\|_{((n-\alpha)q)/n}^{(n-\alpha)/n}. \tag{2.28}$$

From the definition of the Hardy-Littlewood maximal operator it is clear that for arbitrary $y \in \mathbb{R}^n$,

$$M(\chi_{B(x,\delta)}|B(x,\delta)|^{-1})(y)$$

$$\geq \frac{1}{|B(y,2|x-y|)|} \int_{B(y,2|x-y|)} \chi_{B(x,\delta)}(\omega)|B(x,\delta)|^{-1} \, d\omega$$

$$= \frac{\omega_n}{n|x-y|^n}, \tag{2.29}$$

where ω_n is the surface area of the unit sphere in $\mathbb{R}^n$. Since M is bounded on $L_{(n-\alpha)q/n}$, using (2.29), Lemma 2.2 and the equality

$$\int_{B(x,\delta)} \left(\frac{\chi_{B(x,\delta)}}{\|\chi_{B(x,\delta)}\|^{n/(n-\alpha)}} \right)^{(n-\alpha)p/(n(p-1))} = 1,$$

in (2.28), we conclude that

$$\left| \int_{|x-y| \geq \delta} \frac{f(y)}{|x-y|^{n-\alpha}} \, dy \right| \leq 2\|f\|_p \left\| M\left(\chi_{B(x,\delta)}|B(x,\delta)|^{-1}\right) \right\|_{(n-\alpha)q/n}^{(n-\alpha)/n}$$

$$\leq 2\|f\|_p |B(x,\delta)|^{(\alpha-n)/n} \left\| \chi_{B(x,\delta)} \right\|_{(n-\alpha)q/n}^{(n-\alpha)/n}$$

$$\leq 2\|f\|_p |B(x,\delta)|^{(\alpha-n)/n} \left\| \chi_{B(x,\delta)} \right\|_{p/(p-1)}$$

$$\leq 2C\|f\|_p |B(x,\delta)|^{\frac{1}{|B(x,\delta)|} \int_{B(x,\delta)} \left(\frac{\alpha-n}{n} + \frac{p-1}{p}\right)}$$

$$= 2\|f\|_p |B(x,\delta)|^{-1/\tilde{p}_B}, \tag{2.30}$$

where

$$1/\tilde{p}_B = |B(x,\delta)|^{-1} \int_{B(x,\delta)} 1/\tilde{p} = |B(x,\delta)|^{-1} \int_{B(x,\delta)} (1/p - \alpha/n)$$

is the harmonic mean of $\tilde{p} = np/(n - p\alpha)$.

On the other hand,

$$\int_{|x-y|\leq\delta} \frac{|f(y)|}{|x-y|^{n-\alpha}} \, dy \leq \sum_{1}^{\infty} \int_{2^{-k}\delta<|x-y|<\delta 2^{-k+1}} \frac{|f(y)|}{|x-y|^{n-\alpha}} dy$$

$$\leq \sum_{1}^{\infty} (2^{-k}\delta)^{\alpha-n} \int_{2^{-k}\delta\leq|x-y|\leq 2^{-k+1}\delta} |f(y)| \, dy$$

$$\leq c2^n\delta^\alpha \sum_{1}^{\infty} 2^{-k\alpha} \frac{1}{|B(x,2^{-k+1}\delta)|} \int_{|x-y|\leq 2^{-k+1}} |f(y)| \, dy$$

$$\leq c(n)\frac{\delta^\alpha}{2^\alpha - 1} M(f)(x). \tag{2.31}$$

As is apparent from Proposition 1.5, for $\|f\|_p = 1$ there exists $r > 0$ such that

$$\int (Mf(x))^{p(x)} dx \leq C\|Mf\|_p^r \leq C_p^{\prime r} \leq C', \tag{2.32}$$

where C' does not depend on f; also

$$\int_{\mathbb{R}^n} \left(\frac{1}{|B|} \int_B \frac{dy}{(e+|y|)^{n+1}}\right)^{\tilde{p}_-} dx \leq C(n,p) \left\|M\left(\frac{1}{(e+|\cdot|)^{n+1}}\right)\right\|_{\tilde{p}_-}^{\tilde{p}_-}. \tag{2.33}$$

The estimates (2.30) and (2.31) hold in particular for $\delta = (Mf(x))^{-p(x)/n}$. Substitution of this value for δ in inequalities (2.30) and (2.31), an application of Lemma 2.4 and the estimates (2.32) and (2.33) imply that for $\|f\|_p \leq 1$, there exists a positive constant C, depending only on n, p, α, such that

$$\int_{\mathbb{R}^n} \left|\int_{\mathbb{R}^n} \frac{f(y)\chi_{\{y:|x-y|\geq\delta\}}}{|x-y|^{n-\alpha}} \, dy\right|^{\tilde{p}(x)} dx \leq C$$

and

$$\int_{\mathbb{R}^n} \left(\int_{\mathbb{R}^n} \frac{|f(y)|\chi_{\{y:|x-y|\leq\delta\}}}{|x-y|^{n-\alpha}} \, dy\right)^{\tilde{p}(x)} dx \leq C.$$

Adding the preceding inequalities and invoking the convexity of the modular $\rho_{\tilde{p}}$ it is readily derived that for $\|f\|_p \leq 1$, we have

$$\int_{\mathbb{R}^n} |I_\alpha f|^{\tilde{p}(x)} \, dx \leq 2^{\tilde{p}^*-1}C;$$

by definition of the Luxemburg norm, the above estimate shows, for $\|f\|_p \leq 1$, that

$$\|I_\alpha f\|_{\tilde{p}} \leq 2^{\tilde{p}^* - 1} \max\left\{C^{1/p^*}, C^{1/p_*}\right\} = C',$$

from which it easily follows that for all $f \in L_p(\Omega)$,

$$\|I_\alpha f\|_{\tilde{p}} \leq C'(n, p, \alpha)\|f\|_p. \qquad \square$$

2.5   Poincaré-type Inequalities

Poincaré-type inequalities play a central role in the study of differential operators. This section is devoted to the versions that will be utilized in Chapters 3 and 4. We mainly followed the spirit of [23].

2.5.1   *A remark on the geometry of the domain*

Let K be a right circular cone with vertex P and height M. Denote the axis of the cone by a. A straightforward calculation shows that there exist positive constants α, γ, $\sigma > 1$, each depending only on the aperture of the cone, such that given an arbitrary point $x \neq P$ in a, there exist finitely many distinct balls $B_i(x_i, d_i) \subset K$, $1 \leq i \leq k$, k depending on x and satisfying the following conditions:

- $x_1 = x$.

- $B(x_k, d_k)$ contains the centre of the base of the cone K.

- If $j \leq l$, then $B_j \subseteq \sigma B_l$.

- Each point in K belongs to at most γ balls from the family $B_1, ..., B_k$.

- For each $j = 1, 2, ..., k - 1$ there exists a ball $C_j \subseteq B_j \cap B_{j+1}$ whose radius c_j is subject to the estimates

$$\alpha d_j \leq c_j.$$

Theorem 2.9. *Let $\Omega \subset \mathbb{R}^n$ be a bounded Lipschitz domain. Then there exists a ball B with closure contained in Ω, positive constants α, β and a natural number m, all three depending only on Ω, such that given an arbitrary $x \in \Omega$ there exists a finite family of distinct balls $B_1, B_2, ..., B_l = B$, depending on x, with each $B_i \subset \Omega$, for which:*

- $x \in B_1$.

- *If $k \le j$, then $B_k \subseteq \alpha B_j$.*

- *Any point $z \in \Omega$ belongs to at most m of the balls $B_1, ..., B_l$.*

- *For each $j = 1, 2, ..., l-1$ there exists a ball $C_j \subseteq B_j \cap B_{j+1}$ whose diameter satisfies*

$$\alpha \operatorname{diam}(B_j) \le \operatorname{diam}(C_j).$$

Proof. For small enough $\delta > 0$, the compact set

$$\Omega_\delta = \{x \in \Omega : \operatorname{dist}(x, \partial\Omega) \ge \delta\}$$

is covered by a finite family $\mathcal{F}$ of balls $B(x, \varepsilon_x) \subset \Omega$; let R_m and R_M be respectively the minimum and the maximum radii of such balls. The intersection of two balls in $\mathcal{F}$, if non-empty, contains a ball of positive radius. Since there is a finite number of such intersections, the minimum radius r of such balls is positive and we set $\rho = \frac{r}{2}$. Fix a ball $B \in \mathcal{F}$ and $z \in B$. For any $x \in \Omega_\delta$ consider a continuous path

$$\gamma : [0,1] \to \Omega_\delta, \quad \gamma(0) = x, \quad \gamma(1) = z.$$

Let k be the minimum number of balls from $\mathcal{F}$ that cover $\gamma([0,1])$; select a minimal finite family $\mathcal{G}$ consisting of exactly k balls from $\mathcal{F}$ with

$$\gamma([0,1]) \subset \bigcup_{W \in \mathcal{G}} W.$$

Choose $B_1 \in \mathcal{G}$ with $x \in B_1$. If B_1 does not cover $\gamma([0,1])$, there exists $B_2 \in \mathcal{G} \setminus \{B_1\}$ with $B_1 \cap B_2 \ne \emptyset$. Likewise, there exists $B_3 \in \mathcal{G} \setminus \{B_1, B_2\}$ such that $B_2 \cap B_3 \ne \emptyset$. Via this process (which must stop after a finite number of steps) a family of balls $B_1, B_2, ..., B_l$ (l depends only on Ω) is constructed that fulfills the following conditions: $x \in B_1$, $z \in B_l$, $B_j \cap B_{j+1}$ contains a ball C_j with

$$\operatorname{diam}(B_j) \le (R_M/\rho)\,\rho \le (R_M/\rho)\operatorname{diam}(C_j).$$

If δ is small enough, any $w \in \Omega \setminus \Omega_\delta$ belongs to the axis of a right circular cone K_w with vertex $P \in \partial\Omega$, height 2δ, and all cones K_w are congruent, i.e., identical up to rotations and translations. Let M_w be the centre of the base of K_w.

Thus, if $x \in \Omega \setminus \Omega_\delta$ there is a finite sequence of balls $V_1, V_2, ..., V_k$, with $x \in V_1$, $M_x \in V_k$ satisfying the conditions indicated in the discussion at the beginning of the section. This completes the proof. $\square$

2.5.2 *Poincaré's inequalities*

Theorem 2.10. *If $p \in \mathcal{P}^{log}(\Omega)$, there exists a positive constant C, depending on the dimension n, Ω and the log-Hölder modulus of continuity of p, such that for all $u \in \overset{\circ}{W}{}^1_p(\Omega)$,*

$$\|u\|_p \leq C \|\nabla u\|_p. \tag{2.34}$$

Likewise, the inequality

$$\left\| u - \frac{1}{|\Omega|} \int_\Omega u \right\|_p \leq C \|\nabla u\|_p \tag{2.35}$$

holds for any $u \in W^1_p(\Omega)$.

Corollary 2.2. *The original norm $\| \cdot \|_{1,p}$ on $\overset{\circ}{W}{}^1_p(\Omega)$ is equivalent to the norm given by*

$$u \longrightarrow \||\nabla u|\|_p.$$

Proof. The proof resembles that of the classical case for constant exponent p ([48]); the variability of p comes into play when considering the Hardy-Littlewood maximal operator. We will present the salient points of the proofs of (2.34)and (2.35); we also refer the reader to [23]. As is customary, ω_n stands for the surface of the unit sphere in $\mathbb{R}^n$.

For $u \in C^1_0(\Omega)$ (extended by zero to $\mathbb{R}^n$), $\omega \in \mathbb{R}^n$, $|\omega| = 1$ and $x \in \Omega$, we notice that by integrating the equality

$$u(x) = -\int_0^\infty D_s u(x + s\omega) \, ds$$

it follows that

$$\omega_n u(x) = -\int_0^\infty \int_{S^{n-1}} D_s u(x + s\omega) \, d\omega \, ds;$$

thus

$$u(x) = -\int_0^\infty D_s u(x + s\omega) \, ds = -\frac{1}{\omega_n} \int_0^\infty \int_{S^n} \frac{\partial}{\partial s} u(x + s\omega) \, d\omega ds$$

$$= \frac{1}{\omega_n} \int_\Omega \frac{(x_i - y_i) D_i u(y)}{|x - y|^n} \, dy \leq \frac{1}{n\omega_n} I_{\Omega,1}(|\nabla u|)(x).$$

Denote by g the extension of $|\nabla u|$ by zero to $\mathbb{R}^n$. Then

$$
\begin{aligned}
I_{\Omega,1}(|\nabla u|)(x) &= \sum_0^\infty \int_{2^{-k}<|x-y|<2^{-k+1}} \frac{g(y)\,dy}{|x-y|^{n-1}} \\
&\leq \sum_0^\infty (2^{-k})^{1-n} \int_{2^{-k}\leq|x-y|\leq2^{-k+1}} g(y)\,dy \\
&\leq 2^n \sum_0^\infty 2^{-k}2^{(k-1)n} \int_{|x-y|\leq2^{-k+1}} g(y)\,dy \qquad (2.36) \\
&\leq 2^{n+1}M(g)(x).
\end{aligned}
$$

Here, as pointed out in the discussion following Theorem 1.15, M denotes the Hardy-Littlewood maximal operator, which is bounded on $L_p(\Omega)$ under the assumption that $p \in \mathcal{P}^{\log}(\Omega)$ (see the discussion at the end of Chapter 1 and [23], Chapter 3). In conclusion, there exists a positive constant $C = C(n,\Omega,p)$ such that

$$
\|u\|_p \leq C_1\|I_1(|\nabla u|)\|_{p,\Omega} \leq 2^{n+1}\|Mg\|_p \leq C\|\nabla u\|_p. \qquad (2.37)
$$

Using definition of $\overset{\circ}{W}^1_p(\Omega)$, a straightforward density argument shows that the inequality (2.37) holds for any $u \in \overset{\circ}{W}^1_p(\Omega)$.

As to the proof of (2.35) we refer to the notation of Section 2.3.1. The first stage of the proof consists of showing that for any ball $B \in \mathcal{F}$ there exists a positive constant C (depending solely on n and Ω) such that for a.e. $x \in \Omega$, and for every $u \in W^1_1(\Omega)$,

$$
\left| u(x) - \frac{1}{|B|}\int_B u \right| \leq CI_1(|\nabla u|)(x).
$$

For simplicity we will denote the average of the Lebesgue-measurable function u over the Borel set X by

$$
\frac{1}{|X|}\int_X u = \langle u \rangle_X.
$$

First we notice that if Ω is a ball of radius $s/2$ and $u \in C^\infty(\Omega) \cap W^1_1(\Omega)$, then for any $x,y \in \Omega$ (we make no distinction between ∇u and its extension by 0 outside Ω), one has

$$
u(x) - u(y) = -\int_0^{|x-y|} \nabla u\left(x + r\frac{y-x}{|y-x|}\right) \cdot \frac{y-x}{|y-x|}\,dr.
$$

Fixing x and computing the average over Ω of the functions of y on both sides of the preceding inequality it is clear that

$$
\begin{aligned}
|u(x) - \langle u \rangle_\Omega| &\leq \frac{1}{|\Omega|} \int_\Omega \int_0^{|x-y|} \left| \nabla u \left(x + r\frac{y-x}{|y-x|} \right) \right| \, dr \, dy \qquad (2.38) \\
&\leq \frac{C}{s^n} \int_{|x-y| \leq s} \int_0^\infty \left| \nabla u \left(x + r\frac{y-x}{|y-x|} \right) \right| \, dr \, dy \\
&= \frac{C}{s^n} \int_0^s \rho^{n-1} \int_{S^{n-1}} \int_0^\infty |\nabla u(x+rw)| \, dr \, dw \, d\rho \\
&= \frac{C}{n} \int_0^\infty \int_{S^{n-1}} |\nabla u(x+rw)| \, dr \, dw \\
&\leq \frac{C}{n} \int_0^\infty r^{n-1} \int_{S^{n-1}} \frac{|\nabla u(x+rw)|}{r^{n-1}} \, dw \, dr \qquad (2.39) \\
&= \frac{C}{n} \int_\Omega \frac{|\nabla u(y)|}{|x-y|^{n-1}} \, dy.
\end{aligned}
$$

To generalize the proof to the case of a bounded Lipschitz domain Ω, let $x \in \Omega$ and consider the collection of balls $B_1, ..., B_l$ given by Theorem 2.9. Then, for $u \in C^\infty(\Omega) \cap W_1^1(\Omega)$ and $1 \leq j \leq l-1$,

$$
\begin{aligned}
|\langle u \rangle_{B_j} - \langle u \rangle_{C_j}| &\leq \frac{1}{|C_j|} \int_{C_j} |u(y) - \langle u \rangle_{B_j}| \, dy \\
&\leq \frac{C}{\alpha^n |B_j|} \int_{B_j} |u(y) - \langle u \rangle_{B_j}| x \, dy \\
&\leq C(\Omega, n) \, \operatorname{diam}(B_j)^{1-n} \int_{B_j} |\nabla u(y)| \, dy \\
&\leq C(\Omega, n) \int_{B_j} \frac{|\nabla u(y)|}{|x-y|^{n-1}} \, dy \\
&= C(\Omega, n) \int_\Omega \frac{|\nabla u(y)| \chi_{B_j}(y)}{|x-y|^{n-1}} \, dy, \qquad (2.40)
\end{aligned}
$$

since $x \in B_1 \subset \alpha B_j$.
Analogously,

$$
|\langle u \rangle_{B_{j+1}} - \langle u \rangle_{C_j}| \leq C(\Omega, n) \int_\Omega \frac{|\nabla u(y)| \chi_{B_{j+1}}(y)}{|x-y|^{n-1}} \, dy. \qquad (2.41)
$$

Inequalities (2.38), (2.40) and (2.41) give

$$|u(x) - \langle u \rangle_B| \leq |u(x) - \langle u \rangle_{B_1}| + \sum_{j=1}^{l-1} |\langle u \rangle_{B_{j+1}} - \langle u \rangle_{B_j}|$$

$$\leq CI_1(|\nabla u|) + \sum_{j=1}^{l-1} |\langle u \rangle_{B_j} - \langle u \rangle_{C_j}| + \sum_{j=1}^{l-1} |\langle u \rangle_{B_{j+1}} - \langle u \rangle_{C_j}|$$

$$\leq CI_1(|\nabla u|) + 2C(\Omega, n) m \int_\Omega \frac{|\nabla u(y)| \sum_{j=0}^{l-1} \chi_{B_j}(y)}{|x-y|^{n-1}} \, dy$$

$$\leq C \int_\Omega \frac{|\nabla u(y)|}{|x-y|^{n-1}} \, dy.$$

Thus there exists a positive constant $C(\Omega, p)$ for which the estimate

$$|u(x) - \langle u \rangle_B| \leq CI_1(|\nabla u|) \tag{2.42}$$

holds for all $x \in \Omega$ and $u \in W_1^1(\Omega)$. A standard density argument based on Theorem 2.7 shows that (2.42) remains valid for general $u \in W_p^1(\Omega)$. To conclude the proof of (2.35), we observe that on account of Theorem 1.5, for any ball $B \subset \Omega$, we have

$$\|\langle u \rangle_\Omega - \langle u \rangle_B\|_p = |\langle u \rangle_\Omega - \langle u \rangle_B| \|1\|_p$$

$$\leq \frac{1}{|\Omega|} \|u - \langle u \rangle_B\|_{L_1(\Omega)} \|1\|_p$$

$$\leq C(\Omega, p) \frac{1}{|\Omega|} \|1\|_p \|1\|_{p'(\cdot)} \|u - \langle u \rangle_B\|_p.$$

In the light of (2.42) and (2.36), this immediately yields (2.35), since

$$\|u - \langle u \rangle_\Omega\|_p \leq \|u - \langle u \rangle_B\|_p + \|\langle u \rangle_B - \langle u \rangle_\Omega\|_p \tag{2.43}$$

$$\leq C(n, \Omega, p) \|\nabla u\|_p. \qquad \square$$

If $p_+ < n$, the boundedness result for the Riesz Potential given in Theorem 2.8 and the pointwise inequality (2.42) imply the estimate

$$\|u - \langle u \rangle_\Omega\|_{np/(n-p)} \leq C(n, \Omega, p) \|\nabla u\|_p, \tag{2.44}$$

which follows along the same lines as (2.43). Finally we point out that the condition on the exponent can be somewhat relaxed. Out of the various results in this direction (see [23], Section 7.2) we emphasize the following:

Theorem 2.11. *If p is uniformly continuous in a bounded, Lipschitz domain Ω, then there exists a positive constant $C = C(\Omega, p, n)$ such that for every $u \in \overset{\circ}{W}_p^1(\Omega)$, we have*

$$\|u\|_p \leq C \|\nabla u\|_p.$$

2.6   Embeddings

In this section the classical Sobolev embeddings are extended to the variable-integrability case via the potential-theoretic methods developed in the previous section. It is assumed that $p \in \mathcal{P}(\Omega)$ where Ω is a bounded domain with a Lipschitz boundary.

Theorem 2.12. *If $p \in \mathcal{P}^{log}(\Omega)$, $1 < p_- \leq p_+ < n$ and*

$$q : \Omega \to [1, \infty]$$

is Borel-measurable with $q(x) \leq p^*(x) = \frac{np(x)}{n - p(x)}$, the embeddings

$$\overset{\circ}{W}{}^1_p(\Omega) \hookrightarrow L_q(\Omega) \tag{2.45}$$

and

$$W^1_p(\Omega) \hookrightarrow L_q(\Omega) \tag{2.46}$$

hold, with norms depending only on q, n and the Lipschitz character of Ω.

Proof.   The boundedness of the embedding (2.45) follows directly from the pointwise estimate (2.42), the boundedness of the Riesz potential

$$I_{\Omega,1} : L_p(\Omega) \to L_{p^*}(\Omega)$$

and the boundedness of the embedding $L_{p^*}(\Omega) \hookrightarrow L_q(\Omega)$.
As to (2.46), it can be easily verified that if $|\Omega| \geq 1$, then

$$\rho_q \left(\frac{\langle u \rangle_\Omega}{\langle u \rangle_\Omega |\Omega|^{1/q_-}} \right) \leq 1; \tag{2.47}$$

likewise, if $|\Omega| < 1$,

$$\rho_q \left(\frac{\langle u \rangle_\Omega}{\langle u \rangle_\Omega |\Omega|^{1/q_+}} \right) \leq 1. \tag{2.48}$$

Inequalities (2.47) and (2.48) imply by definition of the Luxemburg norm that

$$\|\langle u \rangle_\Omega\|_q \leq \max\{1, |\Omega|^{1/q_+ - 1}\} \|u\|_1. \tag{2.49}$$

The estimate (2.43), inequality (2.49), Hölder's inequality and the estimate (2.44) finally yield

$$\begin{aligned}
\|u\|_q &\leq \|u - \langle u \rangle_\Omega\|_q + \|\langle u \rangle_\Omega\|_q \\
&\leq C(p, n, \Omega) \|u - \langle u \rangle_\Omega\|_{p^*} + C(\Omega, q) \|u\|_p \\
&\leq C(\Omega, n, p) \left(\|u\|_p + \|\nabla u\|_p \right).
\end{aligned}$$

$\square$

Theorem 2.12 remains valid for bounded α-John domains, that is for domains containing a point X_0 with the property that any other point in the domain can be joined to X_0 by means of a rectifiable path γ (parametrized by arc-length) such that for each value t of the parameter, the ball $B(\gamma(t), \frac{t}{\alpha})$ is contained in the domain (see [23] and the references therein).

2.7 Hölder Spaces with Variable Exponents

In direct analogy with the classical definition, if Ω is a bounded domain with a Lipschitz boundary and $\alpha : \Omega \to (0, 1]$ is Borel-measurable, the Hölder space $C^{\alpha(\cdot)}(\Omega)$ (also denoted by $C^{\alpha}(\Omega)$ when there is no danger of misinterpretation) is defined by means of the norm

$$\|f\|_{C^{\alpha(\cdot)}(\Omega)} = \|f\|_{\infty} + \sup_{0 < |x-y| < 1} \frac{|f(x) - f(y)|}{|x - y|^{\max\{\alpha(x), \alpha(y)\}}};$$

these spaces emerge as the natural embedding spaces for variable-integrability Sobolev spaces when the integrability index exceeds the dimension n, as was shown in [34]. Note that these variable-exponent spaces coincide with the classical Hölder spaces when α is constant. The proof presented below is taken from [23].

Theorem 2.13. *For $p \in \mathcal{P}^{\log}(\Omega)$ and $n < p_-$, the embedding*

$$W_p^1(\Omega) \hookrightarrow C^{1-n/p}(\Omega)$$

is continuous; that is, there exists a positive constant $C = C(n, p, \Omega)$ such that for all $u \in W_p^1(\Omega)$, the estimate

$$\|u\|_{\infty,\Omega} + \sup_{0 < |x-y| < 1} \frac{|u(x) - u(y)|}{|x - y|^{1-n/p(x)}} \leq C\|u\|_{1,p}$$

holds.

Proof. From the boundedness of the embeddings

$$W_p^1(\Omega) \hookrightarrow W_{p_-}^1(\Omega) \hookrightarrow L^{\infty}(\Omega)$$

it immediately follows that there exists a positive constant C, depending only on Ω, p and n, such that for arbitrary $u \in W_p^1(\Omega)$,

$$\|u\|_{\infty} \leq C\|u\|_{1,p}.$$

For such u and for $x, y \in \Omega$, $0 < |x - y| < 1$ let r be chosen in such a way that

$$|x - y| < r < \min\{1, 2|x - y|\}$$

and set $B_\Omega = B(x, r) \cap \Omega$. As usual we set $p_{B_\Omega}^- = \inf_{w \in B_\Omega} p(w)$ and note that under the given assumption, $p_{B_\Omega}^- > n$; invoking the classical Sobolev embedding theorem and the obvious continuity of the inclusion

$$W_p^1(B_\Omega) \hookrightarrow W_{p_{B_\Omega}^-}^1(B_\Omega),$$

it is clear that for $u \in W_p^1(\Omega)$, $x \in B_\Omega$, $y \in B_\Omega$ and $0 < |x - y| < 1$ we have the inequality

$$|u(x) - u(y)| \le C(p, n) r^{1 - n/p_{B_\Omega}^-} \|u\|_{1, p_{B_\Omega}^-, B_\Omega}.$$

Let $z \in B_\Omega$ be such that $p(z) = p_{B_\Omega}^-$. Then $|x - z| < r < 1$; consequently

$$|\log r| \le |\log |x - z||.$$

It follows from the latter inequality and the log-Hölder continuity of p that there exists a positive constant C for which

$$r^{1 - \frac{n}{p_{B_\Omega}^-}} \le C r^{1 - \frac{n}{p(x)}}.$$

It is thus apparent from the above and our choice of r that for a certain positive constant, denoted again by $C = C(p, n, \Omega)$, we have

$$|u(x) - u(y)| \le C |x - y|^{1 - n/p(x)} \|u\|_{1, p},$$

as claimed.   □

In [50] Harjulehto and Hästö showed that if $1 \le p_- \le p^+ \le n$, the estimate

$$\|u\|_{L_p^*(\Omega)} \le C(\Omega, n, p) \|\nabla u\|_p.$$

Here, $p^*(x) = np(x)/(n - p(x))$, $n' = n/(n - 1)$ and $L_p^*(\Omega)$ is the Musielak-Orlicz space defined by the function

$$F(x) = \sum_{k=1}^{\lfloor p^*/n' \rfloor} \frac{|x|^{kn'}}{k!} + \frac{1}{\lfloor p^*/n' \rfloor!} |x|^{p^*},$$

where $\lfloor . \rfloor$ represents the integer part of the given quantity, and it is understood that the second term vanishes for $p = n$. Notice that if $p \equiv n$ a.e. in Ω, the above result is nothing more than the classical embedding of $W_n^1(\Omega)$ in the Musielak-Orlicz space $L_{\exp(t^{n'})}(\Omega)$ (see [50]). For $p_- > n$

(see [55], Theorems 3.4), we have, from the classical embedding results for p constant:

Theorem 2.14. *If $p \in \mathcal{P}^{\log}(\Omega)$ and $p_- > n \geq 1$ then the embedding*

$$\overset{\circ}{W}{}_p^1(\Omega) \hookrightarrow C^{1-n/p_-}(\overline{\Omega})$$

is continuous, that is, there exists a positive constant $C = C(n, p, \Omega)$ such that

$$\|u\|_{\infty,\Omega} + \sup_{x \neq y} \frac{|u(x) - u(y)|}{|x - y|^{1-n/p}} \leq C\|u\|_{\overset{\circ}{W}{}_p^1(\Omega)}.$$

Proof. From Theorem 2.13 and the classical results for constant integrability, it follows that

$$\overset{\circ}{W}{}_p^1(\Omega) \hookrightarrow W_p^1(\Omega) \hookrightarrow W_{p_-}^1(\Omega) \hookrightarrow C^{1-n/p_-}(\Omega);$$

in particular, after possible modification on a set of zero Lebesgue measure, every $u \in W_p^1(\Omega)$ is continuous. $\qquad\square$

2.8 Compact Embeddings Revisited

As remarked earlier, the impact of compactness results on partial differential equations is indisputable; fortunately many of the compactness properties of the embeddings survive in the case of variable exponent, as was shown at the beginning of this section. Here we present a potential-theoretic approach to compactness. The novelty in this approach is that, in contrast to those in Section 2.3, no special behaviour of the exponent p is required on the boundary of Ω. Observe that Theorem 2.16 is stronger than Theorem 2.3. In this section the general assumption is that the exponent p belongs to $\mathcal{P}^{\log}(\Omega)$. The reader interested in expanding the results of this section should consult [23] or [46].

Theorem 2.15. *If $p \in \mathcal{P}^{\log}(\Omega)$ is bounded, then*

$$\overset{\circ}{W}{}_p^1(\Omega) \hookrightarrow\hookrightarrow L_p(\Omega).$$

Proof. The boundedness of p yields the existence of a positive constant C, independent of ε and u, such that for all $\varepsilon > 0$ and $u \in W_p^1(\mathbb{R}^n)$,

$$\|\varphi_\varepsilon * u - u\|_p \leq \varepsilon C \|\nabla u\|_p; \tag{2.50}$$

this can be seen by considering $u \in C_0^\infty(\mathbb{R}^n)$ and observing that for arbitrary $x \in \mathbb{R}^n$,

$$u(x - y) - u(x) = \int_0^1 \nabla u(x - ty) \cdot y \, dt;$$

using the definition of the mollifier and changing the order of integration in the above expression, it is readily concluded that

$$\varphi_\varepsilon * u(x) - u(x) = \int_0^1 \int_{\mathbb{R}^n} \varphi_\varepsilon(y) \nabla u(x - yt) \cdot y \, dy \, dt,$$

which after a change of variable in the right-hand side shows that

$$|\varphi_\varepsilon * u(x) - u(x)| \leq \int_0^1 \int_{\mathbb{R}^n} |\varphi_{t\varepsilon}(y) \nabla u(x - y) \cdot t^{-1} y| \, dy \, dt$$

$$= \varepsilon \int_0^1 |\varphi_{t\varepsilon}| * |\nabla u|(x) \, dt. \qquad (2.51)$$

As it can be verified by a simple computation, the integral in (2.51) coincides with the Bochner integral of the integrand, whence the triangle inequality yields

$$\left\| \int_0^1 |\varphi_{t\varepsilon}| * |\nabla u|(x) \, dt \right\|_p \leq \| |\varphi_{t\varepsilon}| * |\nabla u| \|_p \leq C \|\varphi\|_1 \|\nabla u\|_p,$$

from which (2.50) follows at once for smooth u and can be easily extended to $u \in \overset{\circ}{W}_p^1(\Omega)$ utilizing a standard density argument.

By virtue of the Banach-Alaoglu theorem, to verify the compactness of the Sobolev embedding it is sufficient to show that any sequence $(v_k)_{k \in \mathbb{N}} \subset \overset{\circ}{W}_p^1(\Omega)$ which converges weakly to 0 has an $L_p(\Omega)$-strongly convergent subsequence. Notice that for $x \in \Omega$,

$$\varphi_\varepsilon * v_k(x) = \frac{1}{\varepsilon^n} \int_{|x - y| \leq \varepsilon} \varphi \left(\frac{x - y}{\varepsilon} \right) v_k(y) \, dy$$

$$\leq \|v_k\|_p \left\| \frac{1}{\varepsilon^n} \varphi \left(\frac{x - \cdot}{\varepsilon} \right) \chi_{\{z : |x - z| \leq \varepsilon\}}(\cdot) \right\|_{p'}$$

$$\leq C(n) \|v_k\|_p; \qquad (2.52)$$

it follows from $v_k \rightharpoonup 0$ that $\varphi_\varepsilon * v_k \to 0$ pointwise *a.e.*; the estimate (2.52) and Lebesgue's dominated convergence theorem imply

$$\lim_{k \to \infty} \int_{\mathbb{R}^n} |\varphi_\varepsilon * v_k(x)|^{p(x)} = 0.$$

Therefore, in the light of (2.50),

$$\|v_k\|_p \leq \|v_k - v_k * \varphi_\varepsilon\|_p + \|v_k * \varphi_\varepsilon\|_p$$

$$\leq C\varepsilon \|\nabla v_k\|_p + \|v_k * \varphi_\varepsilon\|_p,$$

and clearly this implies $\|v_k\|_p \to 0$ as $k \to \infty$. $\qquad \square$

The condition on p in Theorem 2.15 ensures that the Hardy-Littlewood maximal operator defined by

$$Mf(x) = \sup_{x \in B} \frac{1}{|B|} \int_B |f|$$

(where the supremum is taken over all balls B that contain x and are contained in Ω) is bounded; in fact, the conclusion of Theorem 2.15 remains valid under this more general assumption (see [23], Theorem 4.7.2 and Theorem 7.4.2).

Theorem 2.16. *If $p \in \mathcal{P}^{\log}(\Omega)$ and $p_+ < n$, then*

$$\overset{\circ}{W}^1_p(\Omega) \hookrightarrow\hookrightarrow L_q(\Omega)$$

for any $q \in [1, np/(n-p))$.

Proof. For $q \leq p$ the statement follows directly from Theorem 2.15 by consideration of the continuous inclusion

$$L_p(\Omega) \hookrightarrow L_q(\Omega).$$

Define a variable exponent p^* in Ω by $1/p^*(x) = 1/p(x) - 1/n$ $(x \in \Omega)$; for $p < q < p^*$ in Ω, let

$$\alpha(x) = \frac{1}{2}\left(\frac{p^*(x)(q(x)-1)+q}{q(x)p^*(x)} + 1 \right).$$

Then $0 < \alpha(x) < 1$ and

$$(1-\alpha)(p^*q)/(p^* - q) = 1/2 \text{ in } \Omega.$$

Given a sequence (u_k) in the unit ball of $\overset{\circ}{W}^1_p(\Omega)$, (which without loss of generality can be taken to be weakly convergent to zero), set

$$f_k(x) = |u_k(x)|^{\alpha(x)} \ , \ g_k(x) = |u_k(x)|^{1-\alpha(x)}.$$

Young's inequality, Sobolev's embedding theorem and Proposition 1.5 yield the existence of a positive number β for which the following inequality holds:

$$\|u_k\|_q = \|f_k \cdot g_k\|_q \leq 2\, \|f_k\|_{p^*} \|g_k\|_{\frac{qp^*}{p^*-q}}$$

$$\leq C(\Omega, p) \left(\int_\Omega |u_k(x)|^{1/2} \, dx \right)^\beta$$

$$\leq C(\Omega, p) \|u_k\|_{p^*}^\beta. \tag{2.53}$$

By virtue of Sobolev's embedding Theorem 2.15 the right-hand side of (2.53) tends to 0 as $k \to \infty$. $\qquad \square$

The following compactness result is a consequence of the inclusion

$$\overset{\circ}{W}{}^1_p(\Omega) \subset \overset{\circ}{W}{}^1_{p_-}(\Omega)$$

and the classical Rellich-Kondrachov Theorem. Since it will be used in the sequel, we present its simple proof.

Theorem 2.17. *If $p_- > n$, then*

$$\overset{\circ}{W}{}^1_p(\Omega) \hookrightarrow\hookrightarrow C(\overline{\Omega}).$$

Proof.   The inclusions

$$\overset{\circ}{W}{}^1_p(\Omega) \hookrightarrow \overset{\circ}{W}{}^1_{p_-}(\Omega) \hookrightarrow C(\overline{\Omega})$$

are obvious, since $p_- > n$. $\square$

Notes

A state-of-the-art approach to potential theory and harmonic analysis on spaces of variable integrability is given in [18]. An exhaustive treatment of the theory of Sobolev spaces of variable integrability, in a much more general setting is developed in the already cited work [23]. We have confined ourselves to functions defined on domains with a Lipschitz boundary. Most results in this section, though, are valid on more general setting, for example John Domains (see [23]). Other embedding results for variable-exponent Sobolev spaces can be found in [46]. A sensitive point is that of the uniform convexity of the norm

$$u \longrightarrow \||\nabla u|\|_p$$

in the space $\overset{\circ}{W}{}^1_p(\Omega)$ for $1 < p < 2$; to the best of our knowledge it is not known whether the above norm is uniformly convex. For $p \geq 2$ the above norm is uniformly convex. This assertion follows from the inequalities

$$\left|\frac{a+b}{2}\right|^t + \left|\frac{a-b}{2}\right|^t \leq \frac{1}{2}\left(|a|^t + |b|^t\right)$$

valid for any real number $t : t \geq 2$ and any complex numbers a and b (see Adams' book [4], Lemma 2.27) The relevance of uniform convexity will be clear in the analysis of convergence of the eigenvalues of the modular p-Laplacian undertaken in the next Chapter.

     Sobolev spaces have also been introduced in the context of a general metric space in [49]. Generalizations of the embedding theorems given in this section to the case of Sobolev spaces with variable exponent on spaces of homogeneous type were investigated in [7].

Traces of Sobolev functions have raised considerable interest and are essential in the consideration of boundary value problems. We refer the reader to [24] for a characterization of the trace of $W_p^1(\mathbb{R}^n \times (0, \infty))$.

Chapter 3

The $p(\cdot)$-Laplacian

3.1 Preliminaries

A recollection of a few standard results from nonlinear functional analysis will facilitate the treatment of the $p(\cdot)$-Laplacian presented in this chapter. We continue to use the notation from Chapter 1; in particular, given a topological vector space X, X^* will denote its dual; the notation

$$\langle x, \Lambda \rangle$$

will be used to symbolize the action of $\Lambda \in X^*$ on $x \in X$. Let X, Y be real Banach spaces: all Banach spaces in this chapter will be assumed to be real. An operator

$$T : X \longrightarrow Y$$

is said to be bounded if for any bounded set $M \subset X$, the set $T(M)$ is bounded. Recall from Chapter 1 that in the particular case when $Y = X^*$, the operator T is said to be monotone if the inequality

$$\langle x - y, T(x) - T(y) \rangle \geq 0$$

holds for any $x \in X$, $y \in X$; it is called hemicontinuous if for any fixed $x, y \in X$ the real valued function

$$s \longmapsto \langle y, T(x + sy) \rangle : \mathbb{R} \to \mathbb{R}$$

is continuous. Another class of operators is of particular importance in the sequel:

Definition 3.1. An operator

$$T : X \longrightarrow X^*$$

is said to be of type M if, for any weakly-convergent sequence $(x_n) \subset X$ with $x_n \rightharpoonup x$, the conditions

$$T(x_n) \rightharpoonup f \text{ and } \limsup_{n \to \infty} \langle x_n, T(x_n) \rangle \leq \langle x, f \rangle,$$

imply $T(x) = f$.

Theorem 3.1. *Let X be a reflexive Banach space and*

$$T : X \longrightarrow X^*$$

be hemicontinuous and monotone. Then T is of type M.

Proof. For fixed $y \in X$, (x_n), x and f as in Definition 3.1, the assumed monotonicity of T implies that

$$0 \leq \langle x_n - y, T(x_n) - T(y) \rangle$$

for all n; hence

$$\langle x - y, T(y) \rangle \leq \langle x - y, f \rangle .$$

In particular, for any $z \in X$,

$$\left\langle z, T\left(x - \frac{1}{n}z\right) \right\rangle \leq \langle z, f \rangle ;$$

in conjunction with the hemicontinuity property of T this immediately yields

$$\langle z, T(x) \rangle \leq \langle z, f \rangle$$

for all $z \in X$. This implies $T(x) = f$ as claimed. $\qquad\square$

Lemma 3.1. *If $F : \mathbb{R}^n \longrightarrow \mathbb{R}^n$ is continuous and there exists $\varepsilon > 0$ such that $F(u) \cdot u \geq 0$ for each u such that $\|u\| = \varepsilon$, then F has a zero.*

Proof. Denote the unit ball in $\mathbb{R}^n$ by $\mathbf{B}_n$. Assuming $F(x) \neq 0$ for each $x \in \mathbf{B}_n$, one can define the function

$$\psi : \mathbf{B}_n \longrightarrow \mathbf{B}_n$$

by

$$\psi(x) = -\frac{F(\varepsilon x)}{\|F(\varepsilon x)\|}.$$

This function is easily seen be continuous. Let x_0 be a fixed point of ψ whose existence is guaranteed by Brouwer's Theorem, and notice that $x_0 \neq 0$. Then

$$-\frac{F(\varepsilon x_0) \cdot x_0}{\|F(\varepsilon x_0)\|} = \|x_0\|^2,$$

which is a contradiction. $\qquad\square$

The raison d'être for the battery of results in the previous paragraph becomes apparent in the following surjectivity criterion, which will be used in the treatment of the $p(\cdot)$-Laplacian to be developed in the next section:

Theorem 3.2. *Consider a separable and reflexive Banach space X and let*

$$T : X \longrightarrow X^*$$

be of type M and bounded. If, for some $f \in X^*$ there exists $\varepsilon > 0$ for which

$$\langle x, T(x) \rangle > \langle x, f \rangle \tag{3.1}$$

for every $x \in X$ with $\|x\| > \varepsilon$, then f belongs to the range of T.

Proof. Let $\{x_k : k \in \mathbb{N}\}$ denote a basis for the linear span of a countable, dense subset $Y \subset X$. For each natural number n denote the linear span of the set $\{x_1, x_2, ..., x_n\}$ by Y_n, that is

$$Y_n = \langle \{x_1, x_2, ..., x_n\} \rangle$$

and consider the canonical isomorphism

$$i_n : \mathbb{R}^n \longrightarrow Y_n \ , \ i_n(a_1, ..., a_n) = \sum_1^n a_k x_k.$$

We first claim that for each $n \in \mathbb{N}$ there exists $u_n \in Y_n$ satisfying

$$\langle x_k, T(u_n) \rangle = f(x_k) \text{ for } k = 1, 2, ..., n. \tag{3.2}$$

This is a direct consequence of the assumption on f and of Lemma 3.1, for the function

$$F_n : \mathbb{R}^n \longrightarrow \mathbb{R}^n \ , \ F_n(z) = i_n^* \circ T \circ i_n(z) - i_n^* \circ f$$

satisfies the conditions of Lemma 3.1 and the zero u_n thereby given satisfies the equality (3.2). Consequently, for each n,

$$\langle u_n, T(u_n) \rangle = f(u_n),$$

which by virtue of condition (3.1) implies that $\|u_n\| \leq \varepsilon$. Since T is bounded, the reflexivity of X yields the existence of a subsequence (v_n) of (u_n) such that

$$v_n \rightharpoonup v \in X \ , \ T(v_n) \rightharpoonup f.$$

Since T is of type M, we conclude that $T(v) = f$. $\qquad\qquad\square$

This section is concluded with the proof of an inequality whose importance will be appreciated in the next section:

Lemma 3.2. *For $n \in \mathbb{N}$, $x, y \in \mathbb{R}^n$ and constant p,*

$$\frac{1}{2}\left(|x|^{p-2} - |y|^{p-2}\right)(|x|^2 - |y|^2) + (|x|^{p-2} + |y|^{p-2})|x - y|^2$$
$$= (|x|^{p-2}x - |y|^{p-2}y) \cdot (x - y). \tag{3.3}$$

Proof. This follows by straightforward calculation:

$$(|x|^2 - |y|^2)(|x|^{p-2} - |y|^{p-2}) + |x - y|^2(|x|^{p-2} + |y|^{p-2})$$
$$= (x - y)\left((x + y)(|x|^{p-2} - |y|^{p-2}) + (x - y)(|x|^{p-2} + |y|^{p-2})\right)$$
$$= 2(x - y)((|x|^{p-2}x - |y|^{p-2}y)). \qquad \square$$

3.2   The $p(\cdot)$-Laplacian

In analogy with the constant exponent case, for $u \in \overset{\circ}{W}_p^1(\Omega)$, the Dirichlet p-Laplacian $\Delta_p(u)$ corresponding to a Borel-measurable function

$$p : \Omega \longrightarrow [1, \infty)$$

is defined by

$$\Delta_p(u) = \operatorname{div}\left(|\nabla u(x)|^{p(x)-2}\nabla u(x)\right). \tag{3.4}$$

More specifically Δ_p is the (non-linear) operator

$$\Delta_p : \overset{\circ}{W}_p^1(\Omega) \longrightarrow \left(\overset{\circ}{W}_p^1(\Omega)\right)^*$$

such that for all $u, h \in \overset{\circ}{W}_p^1(\Omega)$,

$$\langle h, \Delta_p u \rangle = -\int_\Omega |\nabla u(x)|^{p(x)-2}\nabla u(x) \cdot \nabla h.$$

A few remarks are in order. The functional

$$F : \overset{\circ}{W}_p^1(\Omega) \longrightarrow [0, \infty) \tag{3.5}$$

defined by

$$F(w) = \int_\Omega \frac{|\nabla w|^{p(x)}}{p(x)}$$

is convex, weakly lower-semicontinuous, Fréchet-differentiable and in fact (see also [19]),

$$F' = \Delta_p. \tag{3.6}$$

For the sake of completeness, a thorough proof of (3.6) will be given in Chapter 4. When p is constant, it is obvious that the functional (3.5) is a multiple of the pth power of the Sobolev norm, namely

$$F(w) = \frac{1}{p}\|\nabla w\|^p.$$

For variable p the consideration of the derivative of the norm as opposed to the derivative of the modular (which is essentially the definition we have adopted for the p-Laplacian) leads to a different differential operator, as will be discussed later.

Lemma 3.3. *Let $\Omega \subset \mathbb{R}^n$ be a bounded, Lipschitz domain and $p : \Omega \to \mathbb{R}$ a Borel-measurable function satisfying $1 < p_- \leq p(x) \leq p_+ < \infty$ a.e. in Ω. Under these conditions, the operator*

$$\Delta_p : \overset{\circ}{W}{}^1_p(\Omega) \longrightarrow \left(\overset{\circ}{W}{}^1_p(\Omega)\right)^*$$

is bounded, hemicontinuous and monotone. In addition, Δ_p is of type M.

Proof. Let $S \subset \overset{\circ}{W}{}^1_p(\Omega)$ be bounded, with

$$\sup\{\|\nabla u\|_p : u \in S\} \leq C.$$

For $u \in S$ and w in the unit ball of $\overset{\circ}{W}{}^1_p(\Omega)$,

$$\langle w, \Delta_p u \rangle = \int_\Omega |\nabla u|^{p(x)-2} \nabla u \cdot \nabla w. \tag{3.7}$$

Taking absolute values in equality (3.7) and invoking the variable exponent form of Hölder's inequality it is clear that

$$\sup\left\{\|\Delta_p u\|_{\left(\overset{\circ}{W}{}^1_p\right)^*} : u \in S\right\} \leq C\left(1 + \frac{1}{p_-} - \frac{1}{p_+}\right),$$

which shows that Δ_p is bounded.

For the proof of hemicontinuity, fix $t \in \mathbb{R}$. It is clear that for

$$|s| < |t| + \frac{1}{2} \quad \text{and} \quad 1 < p(x) \leq 2$$

it holds that

$$|\nabla(u + sv)(x)|^{p(x)-1} \leq |\nabla u(x)|^{p(x)-1} + |s|^{p(x)-1}|\nabla v(x)|^{p(x)-1}, \tag{3.8}$$

whereas for $p(x) > 2$,

$$|\nabla(u + sv)(x)|^{p(x)-1} \leq 2^{p(x)-1}(|\nabla u(x)|^{p(x)-1} + |s|^{p(x)-1}|\nabla v(x)|^{p(x)-1}).$$
(3.9)

On the other hand, it follows by definition that

$$\langle v, \Delta_p(u + sv) \rangle = \int_\Omega |\nabla(u + sv)|^{p(x)-2}\nabla(u + sv) \cdot \nabla v.$$
(3.10)

In view of (3.8) and (3.9) the integrand in (3.10) is bounded above by

$$|\nabla u|^{p(x)-1}|\nabla v| + |s|^{p(x)-1}|\nabla v|^{p(x)}\chi_{\{x:1<p(x)\leq 2\}}$$
$$+ 2^{p(x)-1}\left(|\nabla u|^{p(x)-1}|\nabla v| + |s|^{p(x)-1}|\nabla v|^{p(x)}\right)\chi_{\{p(x)\geq 2\}},$$

which is integrable by virtue of Hölder's inequality. A straightforward application of Lebesgue's dominated convergence theorem yields the hemicontinuity of Δ_p.

Since $\overset{\circ}{W}{}^1_p(\Omega)$ is reflexive and separable, once we have shown that Δ_p is monotone it will follow that it is of type M via Theorem 3.1.

The proof of monotonicity relies ultimately on the identity (3.3). For constant $p \geq 2$ and $x, y \in \mathbb{R}^n$ we have the inequality

$$|x - y|^p = |x - y|^{p-2}|x - y|^2 \leq 2^{p-3}|x - y|^2(|x|^{p-2} + |y|^{p-2})$$

which when combined with the identity 3.3 yields the estimate

$$|x - y|^p \leq 2^{p-2}(|x|^{p-2}x - |y|^{p-2}y) \cdot (x - y),$$
(3.11)

whereas for $1 < p \leq 2$ (with the obvious provision that $x \neq 0$ and $y \neq 0$) we have

$$(p - 1)|x - y|^2(1 + |x| + |y|)^{p-2} \leq (|x|^{p-2}x - |y|^{p-2}y)(x - y),$$
(3.12)

which follows from

$$(|x|^{p-2}x - |y|^{p-2}y)(x - y)$$
$$= (x - y)\int_0^1 \frac{d}{dt}\left((|y + t(x - y)|^{p-2}(y + t(x - y)))\right) dt$$
$$= |x - y|^2 \int_0^1 |y + t(x - y)|^{p-2}$$
$$+ (p - 2)\int_0^1 |y + t(x - y)|^{p-4}\left((y + t(x - y))(x - y)\right)^2$$
$$\leq (p - 1)|x - y|^2 \int_0^1 |y + t(x - y)|^{p-2} \geq (p - 1)|x - y|^2(1 + |x| + |y|)^{p-2}.$$

For fixed u and v in $\overset{\circ}{W}_p^1(\Omega)$, the definition of Δ_p yields

$$\langle u - v, \Delta_p(u) - \Delta_p(v) \rangle = \int_\Omega (|\nabla u|^{p(x)-2}\nabla u - |\nabla v|^{p(x)-2}\nabla v) \cdot \nabla(u - v).$$

$$(3.13)$$

The desired conclusion is obtained by splitting the integral in the right hand side of (3.13) over $\{x : 1 < p(x) \le 2\}$ and $\{x : p(x) \ge 2\}$ and applying the inequalities (3.12) and (3.11). $\square$

Theorem 3.3. *Let $\Omega \subset \mathbb{R}^n$ be a bounded, Lipschitz domain and $p : \Omega \to \mathbb{R}$ a Borel-measurable function satisfying $1 < p_- \le p(x) \le p_+ < \infty$ a.e. in Ω. Then the p-Laplacian is a homeomorphism of $\overset{\circ}{W}_p^1(\Omega)$ onto its dual.*

Proof. The surjectivity of Δ_p is derived from Theorem 3.2, injectivity follows from the Poincaré inequality and inequalities (3.11) and (3.12); and the continuity of $\Delta_{p(\cdot)}^{-1}$ ensues from a functional-analytic argument coupled with inequalities (3.11) and (3.12). We proceed to the details of the proof. Fix $f \in \left(\overset{\circ}{W}_p^1(\Omega)\right)^*$. For $u \in \overset{\circ}{W}_p^1(\Omega)$ with

$$\|\nabla u\|_p > \max\left\{1, \|f\|^{1/(p_--1)}_{\left(\overset{\circ}{W}_p^1(\Omega)\right)^*}\right\},$$

we have

$$1 = \int_\Omega \frac{|\nabla u|^{p(x)}}{\|\nabla u\|_p^{p(x)}} \le \frac{1}{\|\nabla u\|_p^{p_-}} \int_\Omega |\nabla u|^{p(x)};$$

thus for such u,

$$\langle u, \Delta_{p(\cdot)}u \rangle = \int_\Omega |\nabla u|^{p(x)} \ge \|\nabla u\|_p^{p_-} = \|\nabla u\|_p^{p_--1}\|\nabla u\|$$

$$> \|f\|_{\left(\overset{\circ}{W}_p^1(\Omega)\right)^*}\|\nabla u\|_p,$$

which by virtue of Theorem 3.2 implies that f is in the range of Δ_p, hence that the latter is surjective.

The injectivity of the map can be obtained as a consequence of inequalities (3.11) and (3.12) in conjunction with Poincaré inequality. To see this, consider two functions u and $v \in \overset{\circ}{W}_p^1(\Omega)$ with

$$\Delta_p(u) = \Delta_p(v).$$

We estimate the modular of $\nabla u - \nabla v$ as follows:

$$\int_\Omega |\nabla u - \nabla v|^{p(x)} =$$

$$\int_{\{x:p(x)>2\}} |\nabla u - \nabla v|^{p(x)} + \int_{\{x:1<p(x)\leq 2\}} |\nabla u - \nabla v|^{p(x)}. \qquad (3.14)$$

The first integral in (3.14) is, according to (3.11), bounded by

$$2^{p_+-2} \int_\Omega (|\nabla u|^{p(x)-2}\nabla u - |\nabla v|^{p(x)-2}\nabla v) \cdot (\nabla u - \nabla v)$$

$$= 2^{p_+-2} \langle \Delta_p(u) - \Delta_p(v), u - v \rangle = 0. \qquad (3.15)$$

Denote the second integral in (3.14) by I:

$$I = \int_{\{x:1<p(x)<2\}} \frac{|\nabla u - \nabla v|^{p(x)}(1 + |\nabla u| + |\nabla v|)^{p(x)(2-p(x))/2}}{(1 + |\nabla u| + |\nabla v|)^{p(x)(2-p(x))/2}}.$$

Utilizing the modular Hölder inequality the right-hand side in (3.15) can be seen to be bounded above by

$$2 \left\| \frac{|\nabla u - \nabla v|^{p(x)}}{(1 + |\nabla u| + |\nabla v|)^{p(x)(2-p(x))/2}} \chi_{\{x:1<p(x)<2\}} \right\|_{2/p}$$

$$\times \left\| (1 + |\nabla u| + |\nabla v|)^{p(x)(2-p(x))/2} \right\|_{2/(2-p)}.$$

Hence, from the norm-modular inequalities in Proposition (1.5), setting

$$C(u,v) = \left\| (1 + |\nabla u| + |\nabla v|)^{p(x)(2-p(x))/2} \right\|_{2/(2-p)}$$

and

$$J = \int_\Omega \frac{|\nabla u - \nabla v|^2}{(1 + |\nabla u| + |\nabla v|)^{(2-p(x))}},$$

we conclude that

$$I \leq C(u,v) \max \left\{ J^{\frac{p_+}{2}}, J^{\frac{p_-}{2}} \right\}. \qquad (3.16)$$

Estimates (3.16) and (3.12) imply that for some positive constant α,

$$I \leq C(u,v) \left\langle u - v, \Delta_{p(\cdot)}(u) - \Delta_{p(\cdot)}(v) \right\rangle^\alpha = 0. \qquad (3.17)$$

From (3.14), (3.15) and (3.17) it follows that $u - v$ is constant. Poincaré inequality now implies that $u = v$.

It remains to verify the continuity of the inverse operator Δ_p^{-1}: to that effect set

$$\Delta_p = T$$

and suppose that for $(v_n)_n \subset \overset{\circ}{W}\vphantom{W}^1_p(\Omega)$ we have

$$\Delta_p(v_n) \to \Delta_p(u).$$

If the sequence $(v_n)_n$ were unbounded, one could extract from it a subsequence $(u_n)_n$ such that for all n,

$$\|u_n\|_p > 1.$$

In this case, set

$$w_n = 1/\|\nabla u_n\|_p u_n$$

and observe that for arbitrary $\phi \in \overset{\circ}{W}\vphantom{W}^1_p(\Omega)$ with $\|\nabla \phi\|_p \leq 1$, the equality

$$|\langle \phi, T(w_n) \rangle| = \left| \int_\Omega \frac{1}{\|\nabla u_n\|_p^{p(x)-1}} |\nabla u_n|^{p(x)-2} \nabla u_n \cdot \nabla \phi \right|,$$

yields

$$|\langle \phi, T(w_n) \rangle| \leq \frac{1}{\|\nabla u_n\|_p^{p_*-1}} \left| \int_\Omega |\nabla u_n|^{p(x)-2} \nabla u_n \cdot \nabla \phi \right|$$

$$\leq \frac{1}{\|\nabla u_n\|_p^{p_*-1}} \|T(u_n)\|_{(\overset{\circ}{W}\vphantom{W}^1_p(\Omega))^*},$$

which implies that

$$\|T(w_n)\|_{(\overset{\circ}{W}\vphantom{W}^1_p(\Omega))^*} \to 0 \text{ as } n \to \infty. \tag{3.18}$$

On the other hand, it follows automatically by definition that

$$\|T(w_n)\| \geq \langle w_n, T(w_n) \rangle = \int_\Omega |\nabla w_n|^{p(x)} = 1,$$

which contradicts (3.18). Consequently, the sequence (v_n) must be bounded in $\overset{\circ}{W}\vphantom{W}^1_p(\Omega)$.

Next we exploit inequalities (3.11) and (3.12). Observe that

$$\int_\Omega |\nabla v_n - \nabla v|^{p(x)} = \tag{3.19}$$

$$\int_{\{x : x \in \Omega, \, p(x) \geq 2\}} |\nabla v_n - \nabla v|^{p(x)} + \int_{\{x \in \Omega, \, 1 < p(x) < 2\}} |\nabla v_n - \nabla v|^{p(x)}.$$

By inequality (3.11), the first term in (3.19) is bounded by

$$\left| 2^{p_+ - 2} \langle T(v_n) - T(v), v_n - v \rangle \right| \leq 2^{p_+ - 2} \|T(v_n) - T(v)\| \|v_n - v\|, \tag{3.20}$$

and the right hand side above tends to 0 as $n \to \infty$ since $T(v_n) \to 0$ and $(v_n)_n$ is bounded.

The second term in (3.19) can be expressed as

$$
\int_{\substack{\{x:x\in\Omega, \\ 1<p(x)<2\}}} \frac{|\nabla v_n - \nabla v|^{p(x)}}{(1 + |\nabla v| + |\nabla v_n|)^{p(x)(2-p(x))/2}} \, (1 + |\nabla v| + |\nabla v_n|)^{p(x)(2-p(x))/2} ,
$$

which a straightforward application of the generalized Hölder inequality shows to be dominated by

$$
\left\| (1 + |\nabla v| + |\nabla v_n|)^{p(2-p)/2} \right\|_{2/(2-p)} \left\| \frac{|\nabla v_n - \nabla v|^{p(x)}}{(1 + |\nabla v| + |\nabla v_n|)^{p(2-p)/2}} \right\|_{2/p} .
$$

Invoking again the boundedness of the sequence $(v_n)_n$ it is immediate that the left factor above is bounded by a constant independent of n. Set

$$
I_n = \int_\Omega |\nabla(v_n - v)|^2 \, (1 + |\nabla v| + |\nabla v_n|)^{p-2}
$$

Choosing n so large that

$$
\|T(v_n) - T(v)\| < \frac{(p_- - 1)}{\sup_n \|v_n\| + \|v\|},
$$

inequality (3.12) yields $I_n < 1$; consequently, using Proposition 1.5 we conclude that

$$
\left\| \frac{|\nabla v_n - \nabla v|^{p(x)}}{(1 + |\nabla v| + |\nabla v_n|)^{p(2-p)/2}} \right\|_{2/p} \leq \max\left\{ I_n^{p_+/2}, I_n^{p_-/2} \right\} = I_n^{p_-/2}.
$$

A further application of inequality (3.12), for such n, implies

$$
I_n^{\frac{p_-}{2}} \leq \frac{1}{p_- - 1} |\langle v_n - v, T(v_n) - T(v) \rangle|
$$

$$
\leq \frac{1}{p_- - 1} \|T(v_n) - T(v)\| \|v_n - v\|
$$

$$
\leq \frac{1}{p_- - 1} (\sup_n \|v_n\| + \|v\|) \|T(v_n) - T(v)\|. \tag{3.21}
$$

The bounds (3.20) and (3.21) show, according to (3.19), that

$$
\int_\Omega |\nabla v - \nabla v_n|^{p(x)} \to 0 \text{ as } n \to \infty
$$

and hence

$$
\|v_n - v\|_{1,p} \to 0 \text{ as } n \to \infty.
$$

Consequently,

$$
\Delta_p^{-1} : \left(\overset{\circ}{W}_p^1(\Omega) \right)^* \longrightarrow \overset{\circ}{W}_p^1(\Omega)
$$

is continuous. $\qquad \square$

Theorem 3.3 can be rephrased as the following existence and uniqueness Theorem due to Fan-Zhang (see [42]):

Theorem 3.4. *For any* $f \in \left(\overset{\circ}{W}{}^1_p(\Omega) \right)^*$ *there exists a unique solution* $u \in \overset{\circ}{W}{}^1_p(\Omega)$ *to the equation*

$$\Delta_p u = f.$$

3.3 Stability with Respect to Integrability

In this section the behaviour of the solutions to the Dirichlet problem for the p-Laplacian with respect to perturbations of the integrability is discussed. In the sequel $\Omega \subset \mathbb{R}^n$ will stand for a bounded Lipschitz domain and all integrability indexes will be assumed to belong to $C(\overline{\Omega})$ unless otherwise specified.

We fix a non-decreasing sequence

$$(p_k) \subset C(\overline{\Omega})$$

with

$$p \le p_k \to q$$

uniformly for each $k \in \mathbb{N}$.

As in the preceding section, for any integrability index $m \in C(\overline{\Omega})$ we fix the notation

$$m_- = \inf_{\Omega} m$$

and

$$m_+ = \sup_{\Omega} m.$$

We furthermore assume

$$1 < p_- = \inf_{\Omega} p$$

and

$$\sup_{\Omega} q = q_+ < \infty.$$

Consider an arbitrary $f \in \left(\overset{\circ}{W}{}^1_p(\Omega) \right)^*$. For each k, let $u_k \in \overset{\circ}{W}{}^1_{p_k}(\Omega)$,

$u_p \in \overset{\circ}{W}{}^1_p(\Omega)$ and $u_q \in \overset{\circ}{W}{}^1_q(\Omega)$ be, respectively, the unique solutions of

$$\Delta_{p_k}(u_{p_k}) = f \tag{3.22}$$

$$\Delta_p(u_p) = f$$

and

$$\Delta_q(u_q) = f$$

given by Theorem 3.4. The following is a variant of Lemma 4.1 in [33].

Lemma 3.4. *Let* $0 < \varepsilon < 1/e$ *be arbitrary and* N *be large enough to guarantee that* $\|p_i - p_j\|_\infty < \varepsilon$ *if* $i, j > N$, $i \neq j$, *and fix* k, j *such that* $N \leq k \leq j$. *Then, each* u_{p_j} *defined by (3.22) is subject to the estimate*

$$\int_\Omega \frac{|\nabla u_{p_j}|^{p_k}}{p_k} \leq \varepsilon \int_\Omega \frac{1}{p_k} + \varepsilon^{-\varepsilon} \int_\Omega \frac{|\nabla u_{p_j}|^{p_j}}{p_k}.$$

Proof.

$$\int_\Omega \frac{|\nabla u_{p_j}|^{p_k}}{p_k} \leq$$

$$\int_{|\nabla u_{p_j}| < \varepsilon} \frac{|\nabla u_{p_j}|^{p_k}}{p_k} + \int_{\varepsilon \leq |\nabla u_{p_j}| < 1} \frac{|\nabla u_{p_j}|^{p_k}}{p_k} + \int_{|\nabla u_{p_j}| \geq 1} \frac{|\nabla u_{p_j}|^{p_k}}{p_k}. \tag{3.23}$$

A brief computation shows that the first integral on the right-hand side above is less than or equal to

$$\varepsilon \int_\Omega \frac{1}{p_k(x)}.$$

Using the conditions on ε, the second integral on the right-hand side of (3.23) can be dealt with as follows:

$$\int_{\varepsilon \leq |\nabla u_{p_j}| < 1} \frac{|\nabla u_{p_j}|^{p_k - p_j}|\nabla u_{p_j}|^{p_j}}{p_k} \leq \int_{|\nabla u_{p_j}| < 1} \frac{\varepsilon^{-\varepsilon}|\nabla u_{p_j}|^{p_j}}{p_k};$$

it is, on the other hand, clear that the last integral in (3.23) is less than or equal to

$$\int_{|\nabla u_{p_j}| \geq 1} \frac{|\nabla u_{p_j}|^{p_j}}{p_k}.$$

The last three inequalities substituted in (3.23) yield the lemma. $\qquad\square$

We are now ready to inspect the effect of the variation of the integrability exponent on the solution of the Dirichlet problem.

Theorem 3.5. *For u_{p_k} and u_q as in the previous paragraph, the following convergence result holds:*

$$\lim_{k\to\infty} \int_\Omega |\nabla u_{p_k}|^{p_k} = - \int_\Omega f u_q = \int_\Omega |\nabla u_q|^q.$$

Proof. By definition of Δ_{p_k} and from Proposition 1.5 it immediately follows that either $\|\nabla u_{p_k}\|_{p_k} \leq 1$ or

$$\|\nabla u_{p_k}\|_{p_k}^{p_k -} \leq \int_\Omega |\nabla u_{p_k}|^{p_k} = \langle f, u_{p_k}\rangle \leq \|f\|_{\left(\overset{\circ}{W^1_p}(\Omega)\right)^*} \cdot \|\nabla u_{p_k}\|_p.$$

Young's inequality in concert with Lemma 1.3 imply that for any $\delta > 0$,

$$\|\nabla u_{p_k}\|_{p_k}^{p_k -} \leq \left(\left(\frac{\|f\|}{\delta}\right)^{p_k-/(p_k - -1)} \frac{p_k - - 1}{p_k -} + \frac{(\|\nabla u_{p_k}\|_p \delta)^{p_k -}}{p_k -} \right) \tag{3.24}$$

$$\leq \Bigg(\left(\frac{\|f\|}{\delta}\right)^{\frac{p_k -}{p_k - -1}} \frac{p_k - - 1}{p_k -}$$

$$+ \frac{\left(\left(\|p - q\|_\infty^{-\|p-q\|_\infty} + \|p - q\|_\infty |\Omega|\right) \|\nabla u_{p_k}\|_{p_k} \delta\right)^{p_k -}}{p_k -} \Bigg).$$

Since δ is arbitrary, inequality (3.24) shows that the numerical sequence

$$(\|\,|\nabla u_{p_k}|\,\|_{p_k})$$

is bounded. As a consequence of the inequality in Lemma 1.3 one has

$$\|\,|\nabla u_{p_k}|\,\|_p \leq \left(\|p - q\|_\infty^{-\|p-q\|_\infty} + \|p - q\|_\infty |\Omega|\right) \|\nabla u_{p_k}\|_{p_k},$$

whence it is immediate that the sequence (u_{p_k}) is bounded in $\overset{\circ}{W^1_p}(\Omega)$. Denote the weak limit of (u_{p_k}) in $\overset{\circ}{W^1_p}(\Omega)$ by u. It remains to show that $u = u_q$. To that effect, we fix any exponent $r \in C(\overline{\Omega})$ such that $r > p$, and choose k large enough so that $p_k > r$ and the inequality

$$\|q - p_k\|_\infty \leq \|q - r\|_\infty / 2$$

holds on Ω. The sequence $(u_{p_{k+j}})_j$ is bounded in $\overset{\circ}{W^1_r}(\Omega)$, since according to Lemma 1.3,

$$\|\nabla u_{p_{k+j}}\|_r \leq \left(\|p - q\|_\infty^{-\|p-q\|_\infty} + \|p - q\|_\infty |\Omega|\right) \|\nabla u_{p_{k+j}}\|_{p_{k+j}};$$

it is therefore clear that $(u_{p_{k+j}})_j$ is weakly convergent in $\overset{\circ}{W}{}_r^1(\Omega)$: Denote the weak limit by $v \in \overset{\circ}{W}{}_r^1(\Omega)$.

The uniqueness of the limit implies that $u = v$, which yields

$$u \in \bigcap_{p \le r < q} \overset{\circ}{W}{}_r^1(\Omega).$$

Sobolev's embedding theorem allows us to assume without loss of generality that when k tends to infinity,

$$u_{p_k} \longrightarrow u$$

strongly in $L_{p_j}(\Omega)$.

For any fixed k, by virtue of the weak lower-semicontinuity of the functional

$$I : \overset{\circ}{W}{}^1_{p_k}(\Omega) \longrightarrow [0, \infty),$$

$$I(w) = \int_\Omega |\nabla w|^{p_k},$$

u is subject to the condition

$$\int_\Omega |\nabla u|^{p_k} \le \liminf_j \int_\Omega |\nabla u_{p_j}|^{p_k}.$$

On the other hand,

$$\int_\Omega |\nabla u_{p_j}|^{p_k} \le \|p_k - p_j\|_\infty |\Omega| + \|p_k - p_j\|_\infty^{-\|p_j - p_k\|_\infty} \int_\Omega |\nabla u_{p_j}|^{p_j} \quad (3.25)$$

which is bounded independently of j by virtue of (3.24) and by the assumption on the sequence (p_k). Fatou's Lemma yields

$$\int_\Omega \liminf |\nabla u|^{p_k} = \int_\Omega |\nabla u|^q \le \liminf \int_\Omega |\nabla u|^{p_k}. \quad (3.26)$$

In view of (3.25), the right-hand-side in (3.26) is finite, so that $u \in \overset{\circ}{W}{}_q^1(\Omega)$. For a given $\delta > 0$, fix positive numbers θ and γ such that

$$(1 + \theta)(1 + \gamma) < 1 + \frac{\delta}{2}.$$

Since $\varepsilon^{-\varepsilon}$ decreases to 1 as $\varepsilon \to 0^+$, given the assumptions on the sequence (p_k), there exists $M > 0$ such that for $M \le k \le j$ one has

$$\|p_j - p_k\|_\infty < \min\left\{\gamma, \frac{\delta}{2|\Omega|}p_-\right\}, \quad \|p_j - p_k\|_\infty^{-\|p_k - p_j\|_\infty} < 1 + \theta.$$

Notice that for such k and j we have

$$\left\|\frac{p_j}{p_k} - 1\right\|_\infty < \gamma. \tag{3.27}$$

Next, Lemma 3.4 and (3.27) guarantee the validity of the following inequality for $M \le k \le j$:

$$\int_\Omega \frac{|\nabla u_{p_j}|^{p_k}}{p_k}$$

$$\le \|p_k - p_j\|_\infty \int_\Omega \frac{1}{p_k} + \|p_k - p_j\|_\infty^{-\|p_k - p_j\|_\infty} \int_\Omega \frac{|\nabla u_{p_j}|^{p_j}}{p_j}(1 + \gamma)$$

$$\le \frac{\delta}{2} + (1 + \theta)(1 + \gamma) \int_\Omega \frac{|\nabla u_{p_j}|^{p_j}}{p_j}$$

$$\le \frac{\delta}{2} + (1 + \frac{\delta}{2}) \int_\Omega \frac{|\nabla u_{p_j}|^{p_j}}{p_j}. \tag{3.28}$$

For each $M \le k$, the functional

$$I : \overset{\circ}{W}_p^1(\Omega) \longrightarrow [0, \infty),$$

$$I(w) = \int_\Omega \frac{|\nabla w|^{p_k}}{p_k}$$

is weakly lower-semicontinuous; it is easy to see then that:

$$\int_\Omega \frac{|\nabla u|^{p_k}}{p_k} \le \liminf_{j \ge k} \int_\Omega \frac{|\nabla u_{p_j}|^{p_k}}{p_k}.$$

Hence, as easily follows from (3.28),

$$\int_\Omega \frac{|\nabla u|^{p_k}}{p_k} \le \liminf_{j \ge k} \int_\Omega \frac{|\nabla u_{p_j}|^{p_k}}{p_k}$$

$$\le \liminf_{j \ge k} \left(\frac{\delta}{2} + (1 + \frac{\delta}{2}) \int_\Omega \frac{|\nabla u_{p_j}|^{p_j}}{p_j}\right).$$

That is,

$$\int_\Omega \frac{|\nabla u|^{p_k}}{p_k} \le \liminf_{j \ge k} \int_\Omega \frac{|\nabla u_{p_j}|^{p_j}}{p_j}. \tag{3.29}$$

Since $u \in \overset{\circ}{W}_q^1(\Omega)$, Lebesgue's dominated convergence theorem yields

$$\lim_{k \to \infty} \int_\Omega \frac{|\nabla u|^{p_k}}{p_k} = \int_\Omega \frac{|\nabla u|^q}{q}.$$

By virtue of the minimizing property of u_{p_k} it is clear that

$$\int_\Omega \frac{|\nabla u_{p_k}|^{p_k}}{p_k} + \int_\Omega f u_{p_k} \le \int_\Omega \frac{|\nabla u|^{p_k}}{p_k} + \int_\Omega f u,$$

which automatically leads to

$$\limsup_{k\to\infty} \int_\Omega \frac{|\nabla u_{p_k}|^{p_k}}{p_k} \le \limsup_{k\to\infty} \int_\Omega \frac{|\nabla u|^{p_k}}{p_k} = \int_\Omega \frac{|\nabla u|^q}{q}.$$

In conjunction with (3.29) this implies that

$$\int_\Omega \frac{|\nabla u|^q}{q} = \lim_{k\to\infty} \int_\Omega \frac{|\nabla u_{p_k}|^{p_k}}{p_k}. \tag{3.30}$$

Again the minimal character of u_{p_k} yields the inequality

$$\int_\Omega \frac{|\nabla u_{p_k}|^{p_k}}{p_k} + \int_\Omega f u_{p_k} \le \int_\Omega \frac{|\nabla u_q|^{p_k}}{p_k} + \int_\Omega f u_q,$$

valid for each k; passing to the limits as $k \to \infty$, taking into account (3.30) and the fact that $u_q \in \overset{\circ}{W}^1_q(\Omega)$, we obtain

$$\int_\Omega \frac{|\nabla u|^q}{q} + \int_\Omega f u \le \int_\Omega \frac{|\nabla u_q|^q}{q} + \int_\Omega f u_q;$$

that is to say, u minimizes the functional

$$G : W^{1,q}_0(\Omega) \to [0,\infty),$$

$$G(v) = \int_\Omega \frac{|\nabla v|^q}{q} + \int_\Omega f v.$$

In turn, this minimizing property of u implies that

$$\Delta_q(u) = f;$$

and, since $u \in \overset{\circ}{W}^1_q(\Omega)$, Theorem 3.4 yields

$$u = u_q.$$

Hence, by letting $k \to \infty$ in the equality

$$\int_\Omega |\nabla u_{p_k}|^{p_k} = -\int_\Omega f u_{p_k},$$

which holds by definition, it follows at once that

$$\lim_{k\to\infty} \int_\Omega |\nabla u_{p_k}|^{p_k} = -\int_\Omega f u = \int_\Omega |\nabla u_q|^q,$$

as claimed. $\square$

Notes

The study of the solvability of Poisson's equation for the variable p-Laplacian was conducted in [42], where some semilinear related problems are also studied. Other boundary conditions can be considered: The treatment of Poisson's equation with Neumann boundary condition has been undertaken in [79], using mainly variational methods similar to the ones in [42]. For constant exponent p, the problem of stability with respect to integrability was treated by Lindqvist [61].

Stability results in the spirit of Theorem 3.5 were given in [9]. Regularity results for solutions of p-Laplacian like differential equations are well known for constant p (see [57] and [77]); the study of regularity of solutions (in the line of [20], [71] and [68]) of differential equations involving non-standard growth constitutes a major area of research. In 1994, Acerbi and Fusco showed that if the exponent p is piecewise constant on Ω, then the minimizers of the functional

$$\mathcal{F}(u) = \int_\Omega |\nabla u|^p$$

are locally Hölder continuous (see [1] for the details.) For log-Hölder continuous exponent p with $1 < p_- \leq p_+ < \infty$, it was shown by Alkhutov, using the ideas of John and Nirenberg in ([52]) that non-negative $W_p^1(\Omega)$-solutions of

$$\operatorname{div}\left(|\nabla u|^{p-2}\nabla u\right) = 0$$

in a bounded domain Ω satisfy Harnack's inequality in Ω. More precisely, he proved (see [5]) that if u is a positive solution of the p-Laplacian in $\Omega \subseteq \mathbb{R}^n$, then for any ball B_{4R} contained in Ω, the estimate

$$\left(\frac{1}{R^n}\int_{B_{2R}} u^q\right)^{\frac{1}{q}} \leq C(n,p,q,n)\left(\inf_\Omega u + R\right)$$

holds for any $q \in (0,\infty)$; he also proved that any solution of the above problem is locally Hölder continuous in Ω. In [6], the same author dealt with a somewhat more general exponent p.

Utilizing a suitable adaptation of the techniques in [14] and [15], Acerbi and Mingione were able to obtain local, higher integrability estimates for solutions of systems of the form

$$\operatorname{div}\left(a(x,\nabla u)\right) = \operatorname{div}\left(|F|^{p-1}F\right);$$

they proved that under suitable, natural conditions for the vector fields a and F, weak solutions in $W_p^1(\Omega)$ are locally as good as F, more precisely, if $F \in L_q^{\mathrm{loc}}(\Omega)$ with $q > p$, then

$$|\nabla u| \in L_q^{\mathrm{loc}}(\Omega).$$

We refer the reader to [2] for the specifics. In the same line of thought, [17] proved the local Hölder continuity of the gradient of the minimizers of the functional

$$\mathcal{F}(u) = \int_{\mathbb{R}^n} |\nabla u|^p,$$

provided the exponent p is greater than one and Hölder continuous; later, Acerbi and Mingione generalized this result to functionals

$$\mathcal{F} : W^{1,1}_{\text{loc}}(\Omega) \to \mathbb{R} \ , \ \mathcal{F}(u) = \int_\Omega f(x, \nabla u);$$

under suitable conditions on p and f (log-Hölder continuity would do for p), they obtain local Hölder continuity of the gradient of the minimizers (see [2]).

In [40], non-linear variable exponent elliptic equations in divergence form are considered. Global Hölder estimates for the gradient of $W^1_p(\Omega)$-solutions are obtained under further assumptions on $\partial\Omega$ and the coefficients of the equation; both Dirichlet and Neumann boundary conditions are treated. Specifically, the author considers the problem

$$-\text{div}\left(A(x, u, \nabla u)\right) = B(x, u, \nabla u)$$

on a bounded domain Ω, with either Dirichlet or Neumann boundary condition and Hölder continuous exponent p with $1 < p_- \leq p_+ < \infty$. If the vector fields A and B satisfy certain natural ellipticity conditions (for these "variable exponents growth conditions", see [40]), it is shown that if $u \in L_\infty(\Omega) \cap W^1_p(\Omega)$ is a solution to the above problem, then ∇u is locally Hölder continuous; furthermore, if $\partial\Omega$ and the boundary data are Hölder continuous, one has $\nabla u \in C^{1,\alpha}(\overline{\Omega})$.

Chapter 4

Eigenvalues

Chapter 4 is devoted to the analysis of the first eigenvalue of the p-Laplacian and its robustness with respect to the integrability index p. Convergence properties of the first eigenfunctions are also analysed.

4.1 The Derivative of the Modular

For the sake of continuity in the exposition we discuss in detail the following differentiability theorems. Throughout this section it is assumed that Ω is a bounded, Lipschitz domain and that p is a Borel-measurable, real valued function on Ω subject to

$$1 < p_- = \inf_{x \in \Omega} p(x) \leq p \leq p^+ = \sup_{x \in \Omega} p(x) < \infty$$

a.e. in Ω.

Theorem 4.1. *The Fréchet derivative $F' \in L_{p'}(\Omega)$ $(1 < p_- \leq p^+ < \infty)$ of the modular*

$$F : L_p(\Omega) \longrightarrow [0, \infty),$$

$$F(u) = \int_\Omega |u(x)|^{p(x)} \, dx,$$

is given by

$$\langle v, F'_u \rangle = \int_\Omega p(x) |u(x)|^{p(x)-2} u(x) v(x) \, dx \quad (v \in L_p(\Omega)). \qquad (4.1)$$

Proof. If $u \in L_p(\Omega)$, $h \in L_p(\Omega)$, $t \in \mathbb{R}$ and $x \in \Omega$, a direct calculation shows that (if $u(x) \neq 0$),

$$\lim_{t \to 0} \frac{|u(x) + th(x)| - |u(x)|}{t} = \frac{u(x)h(x)}{|u(x)|}. \qquad (4.2)$$

On the other hand, pointwise in Ω, we have, for some $\theta \in (0,1)$,

$$\left| \frac{|u(x) + th(x)|^{p(x)} - |u(x)|^{p(x)}}{t} \right| \leq p(x)\,|h(x)|\,|u(x) + \theta t h(x)|^{p(x)-1}. \quad (4.3)$$

The first observation is that the right hand side of the preceding expression is integrable: this can be verified by means of the variable-exponent version of Hölder's inequality; as to the expression inside absolute values on the left-hand side, it can be written as

$$\frac{1}{t} \int_{|u(x)|}^{|u(x)+th(x)|} p(x)s^{p(x)-1}\,ds \to p(x)|u(x)|^{p(x)-2}u(x)h(x) \quad (4.4)$$

pointwise, by virtue of (4.2) with the understanding that

$$|u(x)|^{p(x)-2}u(x)h(x) = 0 \text{ if } u(x) = 0.$$

In view of the bound (4.3) and of (4.4), an application of Lebesgue's dominated convergence theorem gives

$$\lim_{t\to 0} \int_\Omega \frac{|u(x) + th(x)|^{p(x)} - |u(x)|^{p(x)}}{t} = \int_\Omega p(x)|u(x)|^{p(x)-2}u(x)h(x)\,dx;$$

that is, the right-hand side of (4.1) is the Gâteaux derivative of the modular. As was pointed out in the remark following (1.1), the full proof of Theorem 4.1 will follow from the continuity of the gradient with respect to u. This can be accomplished by the following argument: Set

$$\theta(t) = F\left(u + t\frac{v}{\|v\|} \right) \quad (t > 0)$$

and observe that

$$\theta'(t) = F'_{u+t\frac{v}{\|v\|}} \left(\frac{v}{\|v\|} \right).$$

A direct calculation reveals that for some $t \in (0, \|v\|)$,

$$\left| \frac{F(u+v) - F(u) - F'_u(v)}{\|v\|} \right| = \left| F'_{u+t\frac{v}{\|v\|}} \left(\frac{v}{\|v\|} \right) - F'_u \left(\frac{v}{\|v\|} \right) \right|$$

$$= \left| \int_\Omega \left(\left| u + t\frac{v}{\|v\|} \right|^{p(x)-2} \left(u + t\frac{v}{\|v\|} \right) - |u(x)|^{p(x)-2} u(x) \right) \frac{v}{\|v\|}\,dx \right|$$

$$\leq 2 \int_\Omega \left| \left| u + t\frac{v}{\|v\|} \right|^{p(x)-2} \left(u + t\frac{v}{\|v\|} \right) - |u(x)|^{p(x)-2} u(x) \right|^{\frac{p(x)}{p(x)-1}},$$

and the last integrand tends pointwise to zero as $\|v\| \to 0$. A direct application of Lebesgue's dominated convergence theorem yields

$$\lim_{\|v\|\to 0} \frac{F(u+v) - F(u) - F'_u(v)}{\|v\|} = 0. \qquad \square$$

Corollary 4.1. *Under the assumptions of Theorem 4.1, the Fréchet derivative of the modular*

$$G : \overset{\circ}{W}{}^1_p(\Omega) \longrightarrow [0, \infty),$$

$$G(u) = \int_\Omega |\nabla u(x)|^{p(x)} \, dx,$$

is given by

$$\langle v, G'(u) \rangle = \int_\Omega p(x) |\nabla u(x)|^{p(x)-2} \nabla u(x) \cdot \nabla v(x) \, dx \quad \left(v \in \overset{\circ}{W}{}^1_p(\Omega) \right).$$

Proof. The corollary follows immediately from the chain rule for Fréchet differentiation and by observing that

$$G = F \circ S,$$

where

$$S : \overset{\circ}{W}{}^1_p(\Omega) \longrightarrow L_p(\Omega),$$

$$S(u) = |\nabla u| = \left(\sum_1^n \left(\frac{\partial u}{\partial x_j} \right)^2 \right)^{\frac{1}{2}}$$

and that the Fréchet derivative of S at $u \in \overset{\circ}{W}{}^1_p(\Omega)$ is given by

$$\langle h, S'_u \rangle = \int_\Omega \frac{\nabla u}{|\nabla u|} \nabla u \nabla h \quad \left(h \in \overset{\circ}{W}{}^1_p(\Omega) \right). \qquad \square$$

The following differentiation results will also be needed in the sequel:

Theorem 4.2. *Under the assumptions of Theorem 4.1, the Fréchet derivative of the functional*

$$H : L_p(\Omega) \longrightarrow [0, \infty),$$

$$H(u) = \int_\Omega \frac{|u(x)|^{p(x)}}{p(x)} \, dx, \tag{4.5}$$

is given by

$$\langle v, H'(u) \rangle = \int_\Omega |u(x)|^{p(x)-2} u(x) v(x) \, dx \quad (v \in L_p(\Omega)).$$

Likewise, the Fréchet derivative of

$$I : \overset{\circ}{W}{}^1_p(\Omega) \longrightarrow [0, \infty), \tag{4.6}$$

$$I(u) = \int_\Omega \frac{|\nabla u(x)|^{p(x)}}{p(x)} \, dx,$$

is given by

$$\langle v, I'(u)\rangle = \int_\Omega |\nabla u(x)|^{p(x)-2}\nabla u(x)\nabla v(x)\,dx \quad \left(v \in \overset{\circ}{W}^1_p(\Omega)\right).$$

Proof.   We observe that multiplication by a positive function $c \in L_\infty(\Omega)$, namely

$$c : L_p(\Omega) \to L_p(\Omega),$$

$$u \longmapsto cu,$$

is a linear map from $L_p(\Omega)$ into itself. As such, it is Fréchet differentiable and pointwise equal to its gradient. Since the map (4.5) is the composition of multiplication times the bounded function $1/p^{1/p}$ followed by the map in Theorem 4.1, the result follows automatically from the chain rule. The remaining statement follows as in the proof of Corollary 4.1.   □

4.2   Compactness and Eigenvalues

Consider two Banach spaces X and Y with norms $\|\cdot\|_X$ and $\|\cdot\|_Y$ respectively; assuming that X is reflexive, an elementary functional analytic argument shows that the norm of any compact operator is attained in X, more precisely:

Lemma 4.1. *Given Banach spaces X and Y with X reflexive and a compact linear operator*

$$T : X \to Y,$$

there exists $x_0 \in X$ such that

$$\|T(x_0)\|_Y = \sup_{0 \neq x \in X} \frac{\|T(x)\|_Y}{\|x\|_X}.$$

Proof.   By definition of $\|T\|$ there exists a sequence

$$(x_n) \subset B_X$$

with

$$\|T(x_n)\|_Y \geq \|T\| - \frac{1}{n} \text{ for each } n \in \mathbb{N};$$

from the reflexivity of X, (x_n) can be assumed to be weakly convergent to $x_0 \in B_X$. Invoking the compactness of T, one can extract a subsequence (y_k) of (x_n) such that $(T(y_k))$ converges. A fortiori, $(T(y_k))$ converges weakly to $T(x_0)$, hence the convergence must be strong in Y. As is apparent from the choice of the sequences, it holds for each natural number n that

$$\|T\| < \|T(x_n)\|_Y + \frac{1}{n},$$

whence, from the continuity of the norm one has

$$\|T\| \leq \|T(x_0)\|_Y,$$

from which the lemma follows at once. □

The following section reveals the connection between the eigenvalues of a given compact operator and the maximum problem suggested by the previous Lemma. Recall that a Banach space is called smooth if its norm is Gâteaux-differentiable at every non-zero point. We refer the reader to [12], where this connection was first observed, for further comments.

Theorem 4.3. *Let X, Y be smooth Banach spaces with X reflexive; let D_X and D_Y stand for the gradients of the respective norms. If*

$$T : X \longrightarrow Y$$

is a linear, compact operator (whose transpose is denoted by T^*) and $x_0 \in X\backslash\{0\}$ is maximal in the sense that

$$\|T\| = \frac{\|T(x_0)\|_Y}{\|x_0\|_X},$$

then

$$(T^* D_Y T)(x_0) = \|T\| D_X(x_0). \tag{4.7}$$

Proof. The claim follows by direct differentiation: For fixed $h \in X\backslash\{0\}$ and $t \in (-\|x_0\|_X \|h\|_X^{-1}, \|x_0\|_X \|h\|_X^{-1})$, set

$$F(t) = \frac{\|T(x_0 + th)\|_Y}{\|x_0 + th\|_X}.$$

By assumption F has an absolute maximum at $t = 0$, and the smoothness hypothesis guarantees the differentiability of F; moreover, as shown with a simple calculation,

$$0 = F'(0) = \langle T(h), D_Y(T(x_0)) \rangle \frac{1}{\|x_0\|_X} - \frac{\|T(x_0)\|_Y}{\|x_0\|_X^2} \langle h, D_X(x_0) \rangle,$$

which immediately yields (4.7). □

In the light of Theorem 4.1, on account of the smoothness and reflexivity of the Sobolev space $\overset{\circ}{W}{}_p^1(\Omega)$ and by virtue of the compactness of the Sobolev embedding guaranteed by the assumptions imposed on p and Ω, Lemma 4.1 provides the connection between the norm of the Sobolev embedding and the eigenvalue problem for the classical p-Laplacian for constant p. More specifically, let $p \in (1, \infty)$. Denoting by u_0 a maximal function for the Sobolev embedding

$$E : \overset{\circ}{W}{}_p^1(\Omega) \longrightarrow L^p(\Omega),$$

that is,

$$\|E(u_0)\|_p = \|E\|,$$

a direct application of Lemma 4.1 shows that the equality

$$-\Delta_p(u_0) = \frac{1}{\|E\|} |u_0|^{p-2} u_0$$

holds weakly in $\overset{\circ}{W}{}_p^1(\Omega)$.

4.3   Modular Eigenvalues

As a consequence of the lack of homogeneity in the non-constant case, the p-Laplacian introduced in (3.4) is not the derivative of the norm, but, according to Theorem 4.2, the derivative of the modular-type functional (4.6). This fact somewhat obscures the connection between the eigenvalue problem

$$-\Delta_p(u) = \lambda |u|^{p-2} u$$

and the norm of the Sobolev embedding. In this section, the discussion of the eigenvalue problem in the spirit of Section 4.2 is initiated. As in Section 4.1, $\Omega \subset \mathbb{R}^n$ will stand for a bounded, Lipschitz domain; the restriction $p \in C(\overline{\Omega})$ is imposed on the variable exponent p to ensure the validity of the various embedding theorems to be used in this chapter.

The Sobolev space

$$\overset{\circ}{W}{}_p^1(\Omega),$$

unless otherwise noticed, will be considered to be endowed with the norm

$$u \longmapsto \||\nabla u|\|_p.$$

Theorem 4.4. *Let $p \in C(\overline{\Omega})$ and assume*

$$1 < p_- = \inf_{\Omega} p \le p \le \sup_{\Omega} p = p_+ < \infty.$$

Fix $r > 0$. If $v_0 \in \overset{\circ}{W}{}^1_p(\Omega)$ is a solution of the optimization problem

$$\sup \int_{\Omega} \frac{|u|^{p(x)}}{p(x)} \, dx = \int_{\Omega} \frac{|v_0(x)|^{p(x)}}{p(x)} \, dx, \tag{4.8}$$

where the supremum is taken over all u for which

$$\int_{\Omega} \frac{|\nabla u(x)|^{p(x)}}{p(x)} \, dx = r,$$

then v_0 is a solution of the eigenvalue problem

$$-\text{div}\left(|\nabla v_0|^{p(\cdot)-2}\nabla v_0\right) = \left(\frac{\int_{\Omega} |\nabla v_0(x)|^{p(x)} \, dx}{\int_{\Omega} |v_0(x)|^{p(x)} \, dx}\right)^{-1} |v_0|^{p(\cdot)-2}v_0. \tag{4.9}$$

Proof. Let $H(u) = \rho_p\left(\frac{|u|}{p^{\frac{1}{p}}}\right)$ and $I(u) = \rho_p\left(\frac{|\nabla u|}{p^{\frac{1}{p}}}\right)$ be the functionals introduced in Theorem 4.2; assume problem (4.8) has a solution $v_0 \in \overset{\circ}{W}{}^1_p(\Omega)$. According to Theorem 4.2, we must have

$$\langle v_0, I'_{v_0}\rangle = \int_{\Omega} |\nabla v_0|^{p(x)} \, dx > p_- \int_{\Omega} \frac{|\nabla v_0(x)|}{p(x)} \, dx = r p_- > 0.$$

Since for any $v \in \overset{\circ}{W}{}^1_p(\Omega)$ we can write

$$v = v - \frac{\langle v, I'_{v_0}\rangle}{\langle v_0, I'_{v_0}\rangle} v_0 + \frac{\langle v, I'_{v_0}\rangle}{\langle v_0, I'_{v_0}\rangle} v_0,$$

it is apparent that

$$\overset{\circ}{W}{}^1_p(\Omega) = \ker I'_{v_0} \oplus \langle\{v_0\}\rangle;$$

here $\langle\{v_0\}\rangle$ denotes the linear span of v_0 in $\overset{\circ}{W}{}^1_p(\Omega)$. The function W defined by

$$W : \ker I'_{v_0} \oplus \mathbb{R} \longrightarrow [0, \infty),$$
$$W(h, t) = H(\nabla((1+t)v_0 + h)) - r$$

is continuously differentiable in both variables, $W(0,0) = 0$ and

$$W_t(0,0) = \int_{\Omega} |\nabla v_0(x)|^{p(x)} \, dx > 0.$$

As usual, we denote the corresponding partial derivatives of W by W_h and W_t respectively. By virtue of the implicit function theorem, there exist $\varepsilon > 0$, a neighbourhood of zero $U \subset \overset{\circ}{W}_p^1(\Omega)$ and a differentiable function

$$\varphi : U \cap \ker I'_{v_0} \longrightarrow \mathbb{R}$$

such that $\varphi(0) = 0$,

$$\varphi'(0) = W_h^{-1}\left(-W_t(0, \varphi(0))\right)$$

and that for all $u \in U \cap \ker I'_{v_0}$ we have

$$W(u, \varphi(u)) = H(\nabla\left((1 + \varphi(u))v_0 + u\right)) - r = 0.$$

At this point we claim that $W_h(0,0) = 0$. Indeed, for $w \in \ker I'_{v_0}$ it has been shown in Theorem 4.2 that

$$W(w,0) - W(0,0)$$
$$= H(v_0 + w) - r$$
$$= \int_\Omega \frac{|\nabla(v_0 + w)(x)|^{p(x)}}{p(x)}\, dx - \int_\Omega \frac{|\nabla(v_0)(x)|^{p(x)}}{p(x)}\, dx$$
$$= \int_\Omega |\nabla v_0(x)|^{p(x)-2}\nabla v_0 \nabla w(x)\, dx + \|\nabla w\|_p E(w) = \|\nabla w\|_p E(w),$$

with $E(w) \to 0$ as $\|\nabla w\|_p \to 0$, which settles the claim. Next, for arbitrary $h \in \ker I'_{v_0}$, set

$$\mathbf{h} : (-\varepsilon, \varepsilon) \longrightarrow \overset{\circ}{W}_p^1(\Omega) \ , \quad \mathbf{h}(t) = (v_0 + th + \varphi(th)) \,.$$

Clearly $\mathbf{h}(0) = v_0$, $\mathbf{h}$ is differentiable at 0 since so is φ, and as $\varphi'(0) = 0$ we also have $\mathbf{h}'(0) = h$.

From the discussion above it is apparent that the function

$$S : (-\varepsilon, \varepsilon) \longrightarrow [0, \infty)$$
$$S(t) = H\left(v_0 + th + \varphi(th)\right)$$

is differentiable and has a maximum at $t = 0$; it is therefore obvious that

$$S'(0) = H'(v_0)(h) = 0.$$

Hence

$$\ker\left(I'_{v_0}\right) \subset \ker\left(H'_{v_0}\right).$$

Since H'_{v_0} is not identically zero, it must necessarily hold that

$$\ker\left(I'_{v_0}\right) = \ker\left(H'_{v_0}\right),$$

from which it follows that there must exist a constant $\lambda \in \mathbb{R}$ such that

$$H'_{v_0} = \lambda I'_{v_0}. \tag{4.10}$$

Equality (4.10) holds for every $u \in \overset{\circ}{W}_p^1(\Omega)$; the consideration of the case $u = u_0$ immediately yields (4.9). $\qquad\square$

In preparation for the introduction of the notion of modular eigenvalues it is shown below that the optimization problem (4.8) is indeed solvable. To that effect we start with the following Lemma.

Lemma 4.2. *Let $\Omega \subset \mathbb{R}^n$ be a bounded Lipschitz domain and $p \in C(\overline{\Omega})$ be a real-valued function for which*

$$1 \le p_- = \inf_\Omega p \le p \le \sup_\Omega p = p_+ < \infty$$

in Ω. Then, for any $r > 0$, the modular ball

$$B_r = \left\{ w : w \in \overset{\circ}{W}{}^1_p(\Omega) \wedge \int_\Omega \frac{|\nabla w|^{p(x)}}{p(x)} \, dx \le r \right\}$$

is weakly closed in $\overset{\circ}{W}{}^1_p(\Omega)$.

Proof. From the convexity of the modular it follows that B_r is convex. On the other hand, if $(w_k) \subset B_r$ converges strongly to w in $\overset{\circ}{W}{}^1_p(\Omega)$ and $t \in (0,1]$ is arbitrary, we can write

$$\rho_p\left(\frac{|\nabla w|}{p^{\frac{1}{p}}}\right) - \rho_p\left(\frac{|\nabla w_k|}{p^{\frac{1}{p}}}\right)$$

$$= \rho_p\left(\frac{|\nabla t t^{-1}(w - w_k) + (1-t)(1-t)^{-1}\nabla w_k|}{p^{\frac{1}{p}}}\right) - \rho_p\left(\frac{|\nabla w_k|}{p^{\frac{1}{p}}}\right)$$

$$\le t^{1-p^+} \rho_p\left(\frac{|\nabla w - \nabla w_k|}{p^{\frac{1}{p}}}\right) + \left((1-t)^{1-p^+} - 1\right)\rho_p\left(\frac{|\nabla w_k|}{p^{\frac{1}{p}}}\right). \quad (4.11)$$

Analogously it is readily seen that

$$\rho_p\left(\frac{|\nabla w_k|}{p^{\frac{1}{p}}}\right) - \rho_p\left(\frac{|\nabla w|}{p^{\frac{1}{p}}}\right)$$

$$= \rho_p\left(\frac{|\nabla t t^{-1}(w_k - w) + (1-t)(1-t)^{-1}\nabla w|}{p^{\frac{1}{p}}}\right) - \rho_p\left(\frac{|\nabla w|}{p^{\frac{1}{p}}}\right)$$

$$\le t^{1-p^+} \rho_p\left(\frac{|\nabla w_k - \nabla w|}{p^{\frac{1}{p}}}\right) + \left((1-t)^{1-p^+} - 1\right)\rho_p\left(\frac{|\nabla w|}{p^{\frac{1}{p}}}\right). \quad (4.12)$$

The sequence

$$\left(\rho_p\left(\frac{|\nabla w_k|}{p^{\frac{1}{p}}}\right)\right)_k$$

is bounded as it easily follows from the strong convergence of (w_k). Moreover, by virtue of the inequalities (1.5), the first terms in (4.11) and (4.12) converge to 0 as $k \to \infty$; from the arbitrariness of t we conclude that $w \in B_r$. The lemma follows by invoking the standard fact that in a normed space, any strongly closed convex set is weakly closed. $\qquad\square$

Theorem 4.5. *Let $\Omega \subset \mathbb{R}^n$ be a bounded, Lipschitz domain and consider a real-valued function $p \in C(\overline{\Omega})$ subject to*

$$1 < p_- = \inf_\Omega p \le p(x) \le \sup_\Omega p = p_+ < \infty.$$

Then, for each $r > 0$, there exists

$$v_r \in \overset{\circ}{W}{}_p^1(\Omega)$$

with

$$\int_\Omega \frac{|\nabla v_r|^{p(x)}}{p(x)} \, dx = r$$

for which

$$\int_\Omega \frac{|v_r|^{p(x)}}{p(x)} \, dx = \sup \int_\Omega \frac{|u|^{p(x)}}{p(x)} \, dx = S_p(r), \qquad (4.13)$$

where the supremum is taken over all $u \in \overset{\circ}{W}{}_p^1(\Omega)$ for which $\int_\Omega |\nabla u|^{p(x)}/p(x) \, dx \le r$.

Proof. For $u \in \overset{\circ}{W}{}_p^1(\Omega)$ the inequality

$$\int_\Omega |\nabla u|^{p(x)}/p(x) \, dx \le r,$$

implies, in view of Proposition 1.5, that

$$\||\nabla u|\|_p \le \max \left\{ (p_+ r)^{1/p_+}, (p_+ r)^{1/p_-} \right\}. \qquad (4.14)$$

On account of the boundedness of the Sobolev embedding

$$E : \overset{\circ}{W}{}_p^1(\Omega) \to L_p(\Omega)$$

(whose operator norm is denoted by $\|E\|$), we readily obtain

$$\|u\|_p \le \|E\| \, \||\nabla u|\|_p$$
$$\le \|E\| \, ;$$

a further application of Proposition 1.5 yields

$$\rho_p(u) \le \max \left\{ \|E\|^{p+} p_+ r, \|E\|^{p+} (p_+ r)^{\frac{p_+}{p_-}}, \|E\|^{p-} (p_+ r), \|E\|^{p-} (p_+ r)^{\frac{p_-}{p_+}} \right\}.$$

Denoting the right-hand side of the above equality by M_r, it is clear then that for $r > 0$ and $u \in \overset{\circ}{W}{}^1_p(\Omega)$ with $\rho_p \left(\frac{\nabla u}{p^{\frac{1}{p}}} \right) \le r$, we have the estimate

$$\rho_p \left(\frac{u}{p^{\frac{1}{p}}} \right) \le \frac{1}{p_-} M_r.$$

Next, a maximizing sequence is chosen in the usual way: for each $n \in \mathbb{N}$ select $u_n \in \overset{\circ}{W}{}^1_p(\Omega)$ such that

$$\rho_p \left(\frac{\nabla u_n}{p^{\frac{1}{p}}} \right) \le r$$

and

$$S_p(r) - \frac{1}{n} < \rho_p \left(\frac{u_n}{p^{\frac{1}{p}}} \right) \le S_p(r).$$

Inequality (4.14) implies that the sequence (u_n) is bounded in $\overset{\circ}{W}{}^1_p(\Omega)$. By virtue of the reflexivity of $\overset{\circ}{W}{}^1_p(\Omega)$ and the compactness of E it follows that the maximizing sequence $(u_n) \subset \overset{\circ}{W}{}^1_p(\Omega)$ has a subsequence (still denoted by (u_n)) which is strongly convergent in $L_p(\Omega)$ to $u_p \in \overset{\circ}{W}{}^1_p(\Omega)$. On account of Lemma 4.2 we see that

$$\rho_p \left(\frac{|\nabla u_p|}{p^{\frac{1}{p}}} \right) \le r.$$

The identities

$$\frac{u_n}{p^{\frac{1}{p}}} = t t^{-1} \frac{u_n - u_p}{p^{\frac{1}{p}}} + (1-t)(1-t)^{-1} \frac{u_p}{p^{\frac{1}{p}}},$$

and

$$\frac{u_p}{p^{\frac{1}{p}}} = t t^{-1} \frac{u_p - u_n}{p^{\frac{1}{p}}} + (1-t)(1-t)^{-1} \frac{u_n}{p^{\frac{1}{p}}},$$

valid for arbitrary $t \in (0,1)$, together with the convexity of the modular ρ_p yield the two inequalities

$$\rho_p \left(\frac{u_n}{p^{\frac{1}{p}}} \right) - \rho_p \left(\frac{u_p}{p^{\frac{1}{p}}} \right)$$

$$\le t^{1-p^+} \rho_p \left(\frac{u_n - u_p}{p^{\frac{1}{p}}} \right) + \left((1-t)^{1-p^+} - 1 \right) \rho_p \left(\frac{u_p}{p^{\frac{1}{p}}} \right), \qquad (4.15)$$

and

$$\rho_p\left(\frac{u_p}{p^{\frac{1}{p}}}\right) - \rho_p\left(\frac{u_n}{p^{\frac{1}{p}}}\right)$$

$$\leq t^{1-p^+}\rho_p\left(\frac{u_n - u_p}{p^{\frac{1}{p}}}\right) + \left((1-t)^{1-p^+} - 1\right)\rho_p\left(\frac{u_n}{p^{\frac{1}{p}}}\right). \qquad (4.16)$$

Since $u_n \to u_p$ in $L_p(\Omega)$ as $n \to \infty$, by virtue of the inequalities (1.5) the first term in the right-hand side of (4.15) tends to 0 as $n \to \infty$; moreover, it is easily seen from the convergence of (u_n) that the sequence

$$\rho_p\left(\frac{u_n}{p^{\frac{1}{p}}}\right)$$

is bounded, whence the arbitrariness of t yields

$$\rho_p\left(\frac{u_p}{p^{\frac{1}{p}}}\right) = S_p(r). \qquad (4.17)$$

In fact, the maximizing function u_p satisfies the condition

$$\rho_p\left(\frac{\nabla u_p}{p^{\frac{1}{p}}}\right) = r;$$

that is, the maximum of the functional ρ_p on the modular ball

$$\rho_p\left(\frac{\nabla u}{p^{\frac{1}{p}}}\right) \leq r$$

is attained on the modular sphere $\rho_p\left(\frac{|\nabla u|}{p^{\frac{1}{p}}}\right) = r$. This last assertion follows by observing that for any $t > 1$, we have

$$\rho_p\left(t\frac{u_p}{p^{\frac{1}{p}}}\right) \geq t^{p^-}\rho_p\left(\frac{u_p}{p^{\frac{1}{p}}}\right) > \sup \int_\Omega \frac{|u|^p}{p}, \qquad (4.18)$$

where the supremum is taken over the modular ball

$$\rho_p\left(|\nabla u|/p^{1/p}\right) \leq r.$$

Hence the inequality

$$\rho_p\left(\frac{\nabla u_p}{p^{\frac{1}{p}}}\right) = \int_\Omega \left|\frac{\nabla u_p}{p}\right|^p dx < r,$$

with

$$t = \left(\frac{r}{\rho_p \left(\frac{\nabla u_p}{p^{\frac{1}{p}}} \right)} \right)^{\frac{1}{p_+}}$$

would imply

$$\rho_p \left(\frac{\nabla t u_p}{p^{\frac{1}{p}}} \right) \leq t^{p_+} \rho_p \left(\frac{\nabla u_p}{p^{\frac{1}{p}}} \right) \leq r,$$

which together with (4.18) contradicts (4.17); in all:

$$\rho_p \left(\nabla u_p / p^{1/p} \right) = r,$$

as claimed. $\qquad\qquad\qquad\qquad\qquad\qquad\qquad\qquad\qquad\qquad\qquad\qquad\square$

Theorem 4.6. *Let the real-valued function* $p \in C(\overline{\Omega})$ *satisfy*

$$1 < p_- \leq p_+ < \infty.$$

Then for every $s > 0$ *there exists* $u_0 \in \overset{\circ}{W^1_p}(\Omega)$ *such that*

$$\int_\Omega \frac{|\nabla u_0|^p}{p} \, dx = \min_{\int_\Omega \frac{|u|^p}{p} \, dx \geq s} \int_\Omega \frac{|\nabla u|^p}{p} \, dx =: I_p(s).$$

Moreover, $\int_\Omega \frac{|u_0|^p}{p} = s.$

Proof. Poincaré's inequality and Proposition 1.5 yield that for $u \in \overset{\circ}{W^1_p}(\Omega)$,

$$\min \left\{ \left(p_- \int_\Omega \frac{|u|^p}{p} \right)^{1/p_-}, \left(p_- \int_\Omega \frac{|u|^p}{p} \right)^{1/p_+} \right\}$$

$$\leq \min \left\{ \left(\int_\Omega |u|^p \right)^{1/p_-}, \left(\int_\Omega |u|^p \right)^{1/p_+} \right\}$$

$$\leq \|u\|_p \leq C \|\nabla u\|_p$$

and that

$$\|\nabla u\|_p \leq \max \left\{ (p_+)^{1/p_+} \left(\int_\Omega \frac{|\nabla u|^p}{p} \right)^{1/p_+}, (p_+)^{1/p_-} \left(\int_\Omega \frac{|\nabla u|^p}{p} \right)^{1/p_-} \right\}.$$

These inequalities show that for $r > 0$ we have

$$I_p(s) := \inf_{\int_\Omega \frac{|u|^p}{p} \geq s} \int_\Omega \frac{|\nabla u|^p}{p} > 0.$$

For each natural number n let $u_n \in \overset{\circ}{W}{}^1_p(\Omega)$ satisfy

$$\int_\Omega \frac{|u_n|^p}{p} \geq s,$$

$$\int_\Omega \frac{|\nabla u_n|^p}{p} < I_p(s) + \frac{1}{n}.$$

Invoking Proposition 1.5 it is easily derived that

$$\|\nabla u_n\|_p \leq \max \left\{ \left(p_+ \int_\Omega |\nabla u_n|^p \right)^{1/p_-} , \left(p_+ \int_\Omega |\nabla u_n|^p \right)^{1/p_+} \right\}$$

$$\leq \max \left\{ (p_+(I_p(s)+1))^{1/p_-} , (p_+(I_p(s)+1))^{1/p_+} \right\};$$

consequently the sequence (u_n) is bounded in $\overset{\circ}{W}{}^1_p(\Omega)$ and can, without loss of generality, be assumed to be weakly convergent. Let u_0 be its weak limit. By virtue of the Sobolev embedding theorem, necessarily

$$u_n \to u_0 \text{ strongly in } L_p(\Omega);$$

inequalities (4.15) and (4.16) yield

$$\rho_p \left(\frac{u_0}{p^{\frac{1}{p}}} \right) \geq s.$$

The weak lower-semicontinuity of the functional

$$u \longmapsto \int_\Omega \frac{|\nabla u|^p}{p},$$

implies, in particular, that

$$\int_\Omega \frac{|\nabla u_0|^p}{p} \leq \liminf \int_\Omega \frac{|\nabla u_n|^p}{p} = I_p(s),$$

from which it is clear that u_0 is indeed a minimizer of the Dirichlet energy integral. Finally, in the eventuality that $\int_\Omega \frac{|u_0|^p}{p} > s$, the inequality

$$t : \left(\frac{s}{\int_\Omega \frac{|u_0|^p}{p}} \right)^{1/p_+} < t < 1,$$

would imply

$$s < t^{p_+} \int_\Omega \frac{|u_0|^p}{p} < \int_\Omega \frac{|t u_0|^p}{p}$$

and

$$\int_\Omega \frac{|\nabla u_0|^p}{p} > t^{p-}\int_\Omega \frac{|\nabla u_0|^p}{p} > \int_\Omega \frac{t^p|\nabla u_0|^p}{p} = \int_\Omega \frac{|\nabla t u_0|^p}{p}.$$

The last two inequalities contradict the definition of u_0, so that necessarily

$$\int_\Omega \frac{|u_0|^p}{p} = s,$$

as claimed. $\qquad\square$

For $r > 0$ let u_r be the maximal function obtained in Theorem 4.5; let

$$\int_\Omega \frac{|u_r|^p}{p} = S_p(r).$$

As a byproduct of the preceding discussion we have the following result:

Corollary 4.2. *For $r > 0$, any maximal function v_r in the sense of Theorem 4.5 satisfies the condition*

$$\int_\Omega \frac{|v_r|^p}{p} = S_p(r) \quad and \quad \int_\Omega \frac{|\nabla v_r|^p}{p} = I_p(S_p(r)). \qquad (4.19)$$

Proof. Obviously $I_p(s) \le r$ and the assumption $I_p(s) < r$ would contradict Theorem 4.5. $\qquad\square$

Therefore, in the notation of the two preceding Theorems, we can write

$$\int_\Omega \frac{|u_r|^p}{p} = S_p(r) = \sup_{\int_\Omega \frac{|\nabla v|^p}{p} \le r} \int_\Omega \frac{|v|^p}{p}$$

and

$$\int_\Omega \frac{|\nabla u_r|^p}{p} = r = \inf_{\int_\Omega \frac{|v|^p}{p} \ge S_p(r)} \int_\Omega \frac{|\nabla v|^p}{p}.$$

For each $r > 0$ denote by $V_{p,r}$ the set of all maximal functions in the modular ball B_r defined in Lemma 4.2, that is,

$$V_{p,r} = \left\{ \omega \in \overset{\circ}{W}{}_p^1(\Omega) : \int_\Omega \frac{|\omega|^p}{p} = \sup_{v \in B_r} \int_\Omega \frac{|v|^p}{p} \right\}. \qquad (4.20)$$

Occasionally, if there is no room for confusion, the subscript p will be dropped. Because of the previous discussion one can assume that any such function ω satisfies (4.19).

As proved in Theorem 4.5,

$$v \in V_r \Rightarrow \int_\Omega \frac{|\nabla v|^p}{p} = r;$$

moreover, for each $u \in V_r$ there exists an eigenvalue $\lambda > 0$ such that

$$-\text{div}\left(|\nabla u|^{p-2}\nabla u\right) = \lambda|u|^{p-2}u;$$

the set $\mathcal{E}_r$ of all such eigenvalues is bounded below by a positive constant. Indeed,

$$m = \sup \int_\Omega |v|^p$$

where the supremum is taken over all v for which

$$\int_\Omega |v|^p \le p^+ r,$$

a simple computation shows that each $\lambda \in \mathcal{E}$ is subject to the inequality

$$\lambda = \frac{\int_\Omega |\nabla u|^p}{\int_\Omega |u|^p} \ge \frac{p_- r}{m} > 0.$$

Let $0 < \lambda_0 = \inf \mathcal{E}$.

Lemma 4.3. *Under the assumptions of Theorem* 4.5, *for each* $r > 0$, *there exists* $w \in V_{p,r}$ *such that*

$$\frac{\int_\Omega |\nabla w|^p}{\int_\Omega |w|^p} = \inf_{u \in V_{p,r}} \frac{\int_\Omega |\nabla u|^p}{\int_\Omega |u|^p}.$$

Proof.   Consider a sequence $(w_n) \subset V_{p,r}$ chosen in such a way that

$$\lambda_0 \le \frac{\int_\Omega |\nabla w_n|^p}{\int_\Omega |w_n|^p} < \lambda_0 + \frac{1}{n};$$

as before, it can be supposed without loss of generality that

$$w_n \rightharpoonup w \in \overset{\circ}{W}{}^1_p(\Omega).$$

The set

$$\left\{ \phi \in \overset{\circ}{W}{}^1_p(\Omega) : \int_\Omega \frac{|\nabla \phi|^p}{p}\, dx \le r \right\}$$

is convex and strongly closed in $\overset{\circ}{W}{}^1_p(\Omega)$; hence it is also weakly closed and therefore

$$\int_\Omega \frac{|\nabla w|^p}{p} \le r.$$

By virtue of the compactness of the Sobolev embedding, it can be assumed without loss of generality that

$$w_n \to w$$

strongly in $L_p(\Omega)$ and thus

$$\int_\Omega |w_n|^p \, dx \to \int_\Omega |w|^p \, dx.$$

Weak lower semi-continuity yields

$$\int_\Omega |\nabla w|^p \, dx \le \liminf_n \int_\Omega |\nabla w_n|^p \, dx$$

$$= \liminf_n \frac{\int_\Omega |\nabla w_n|^p \, dx}{\int_\Omega |w_n|^p \, dx} \int_\Omega |w_n|^p \, dx$$

$$= \lambda_0 \int_\Omega |w|^p \, dx,$$

and thus

$$\frac{\int_\Omega |\nabla w|^p \, dx}{\int_\Omega |w|^p \, dx} = \lambda_0.$$

On the other hand, for each $\varepsilon > 0$, we have for sufficiently large $n \in \mathbb{N}$,

$$\int_\Omega \frac{|w|^p}{p} \, dx > \int_\Omega \frac{|w_n|^p}{p} \, dx - \varepsilon = M_r - \varepsilon,$$

that is, $w \in V_{p,r}$. $\qquad\square$

We make the following remark. If $r > 0$ and p is constant, let v_r be the maximal function given by (4.13) and set

$$v_0 = \frac{v_r}{(rp)^{\frac{1}{p}}};$$

elementary calculations show that $\|\nabla v_0\|_p = 1$; on the other hand, for arbitrary $u \in \overset{\circ}{W}{}_p^1(\Omega), u \ne 0$,

$$v := (pr)^{1/p} \frac{u}{\|\nabla u\|_p}$$

satisfies

$$\int_\Omega \frac{|\nabla v|^p}{p} = r,$$

which, taking into account the maximal character of v_r, implies the inequality

$$\int_\Omega \frac{1}{p} \left| \frac{(pr)^{\frac{1}{p}} u}{\|\nabla u\|_p} \right|^p \, dx \le \int_\Omega \frac{|v_r|^p}{p} \, dx.$$

In turn, the latter inequality leads to

$$\frac{\|\nabla v_0\|_p^p}{\|v_0\|_p^p} = \frac{1}{\int_\Omega \frac{|v_r|^p}{rp}\, dx} = \frac{r}{\int_\Omega \frac{|v_r|^p}{p}\, dx} \leq \frac{r}{\int_\Omega \frac{1}{p}\left|\frac{(pr)^{1/p}u}{\|\nabla u\|_p}\right|^p dx}.$$

From the arbitrariness of u it follows that

$$\frac{\|\nabla v_0\|_p^p}{\|v_0\|_p^p} = \inf_{0\neq u\in W_0^{1,p}(\Omega)} \frac{\|\nabla u\|_p^p}{\|u\|_p^p}.$$

In conclusion, v_0 is *the* eigenfunction corresponding to the first eigenvalue of the p-Laplacian for a constant p.

The last paragraph justifies the following definition:

Definition 4.1. For p and Ω subject to the conditions of Theorem 4.5 and $r > 0$, any of the functions w given by Lemma 4.3 is said to be an eigenfunction corresponding to the (first) eigenvalue

$$\frac{\int_\Omega |\nabla w|^p}{\int_\Omega |w|^p} = \lambda_{p,r} > 0.$$

4.4   Stability with Respect to the Exponent

We now turn our attention to the analysis of the effect of perturbations of the integrability exponent on the modular eigenvalues defined in the foregoing section.

Lemma 4.4. *Let $\Omega \subset \mathbb{R}^n$ be a bounded Lipschitz domain and $p \in C(\overline{\Omega})$ be such that*

$$1 < p_- = \inf_\Omega p \leq p \leq \sup_\Omega p = p_+ < \infty;$$

for each $r > 0$ write

$$S_p(r) = \sup \rho_p\left(\frac{u}{p^{1/p}}\right),$$

where the supremum is taken over all $u \in \overset{\circ}{W}{}_p^1(\Omega)$ for which $\rho_p\left(|\nabla u|/p^{1/p}\right) \leq r$ (see Theorem 4.5). Then for all $\varepsilon > 0$,

$$S_p(r) \leq S_p(r + \varepsilon) \leq \left(\frac{r + \varepsilon}{r}\right)^{p_+} S_p(r). \tag{4.21}$$

Proof. The first inequality in (4.21) is obvious. The second one follows from the observation that for $\rho_p\left(\frac{|\nabla u|}{p^{\frac{1}{p}}}\right) \leq r + \varepsilon$ we have

$$\rho_p\left(\frac{r}{r+\varepsilon}\frac{|\nabla u|}{p^{\frac{1}{p}}}\right) \leq \frac{r}{r+\varepsilon}\rho_p\left(\frac{|\nabla u|}{p^{\frac{1}{p}}}\right) \leq r;$$

consequently,

$$\left(\frac{r}{(r+\varepsilon)}\right)^{p_+}\rho_p\left(\frac{u}{p^{\frac{1}{p}}}\right) \leq S_p(r).$$

In all,

$$S_p(r+\varepsilon) \leq \left(\frac{r+\varepsilon}{r}\right)^{p_+}S_p(r),$$

as claimed. □

Lemma 4.5. *For a bounded domain* $\Omega \subset \mathbb{R}^n$, *a function* $u \in L_\infty(\Omega)$ *and* $p \in C(\overline{\Omega})$ *and* $q \in C(\overline{\Omega})$ *such that for some* $\varepsilon > 0$

$$1 < p_- \leq p(x) \leq q(x) \leq p(x) + \varepsilon < p^+ + \varepsilon < \infty \ , in \ \Omega,$$

we have the inequality:

$$\int_\Omega |u|^q \, dx \leq \left(\|u\|_{L_\infty(\Omega)} + 1\right)^\varepsilon \int_\Omega |u|^p \, dx.$$

Proof. Setting $\Omega_n = \{x \in \Omega : n - 1 \leq |u(x)| < n\}$ one has

$$\int_\Omega |u|^q \, dx = \sum_{n=1}^{\lfloor\|u\|_{L_\infty}+1\rfloor} \int_{\Omega_n} |u|^q \, dx$$

$$= \sum_{n=1}^{\lfloor\|u\|_{L_\infty}+1\rfloor} \int_{\Omega_n} \left|\frac{u}{n}\right|^q n^{q-p}n^p \, dx$$

$$\leq \sum_{n=1}^{\lfloor\|u\|_{L_\infty}+1\rfloor} \int_{\Omega_n} \left|\frac{u}{n}\right|^p n^\varepsilon n^p$$

$$\leq [\|u\|_{L_\infty} + 1]^\varepsilon \int_\Omega |u|^p \, dx.$$

□

The next lemma is a variant of Lemma 1.3.

Lemma 4.6. *Let* $\Omega \subset \mathbb{R}^n$ *be a bounded Lipschitz domain,* $\varepsilon > 0$, *and* $p, q \in \mathcal{P}(\Omega)$ *with* $p < q < p + \varepsilon$ *a.e. in* Ω; *let* $f : \Omega \to \mathbb{R}$ *be Borel-measurable. Then*

$$\int_\Omega |f|^p \, dx \leq \varepsilon|\Omega| + \varepsilon^{-\varepsilon}\int_\Omega |f|^q \, dx. \tag{4.22}$$

Proof.   As in Lemma 1.3, we write

$$\int_\Omega |f|^p\, dx =$$

$$\int_{\{|f|<\varepsilon\}} |f|^p\, dx + \int_{\{\varepsilon<|f|<1\}} |f|^p\, dx + \int_{\{|f|>1\}} |f|^p dx,$$

which leads to the bound

$$\int_\Omega |f|^p\, dx \leq$$

$$\varepsilon|\Omega| + \int_{\{\varepsilon<|f|\leq1\}} |f|^q|f|^{p-q}\, dx + \int_{\{|f|>1\}} |f|^q\, dx,$$

from which the claim is easily derived.   □

Theorem 4.7. *Under the assumptions of Theorem 4.5 and assuming the exponent p satsfies*

$$\inf_\Omega p = p_- > n + \delta$$

for some $\delta > 0$, the first eigenvalue (Definition 4.1) of the modular p-Laplacian is exponent-stable. More specifically, given a bounded domain $\Omega \subset \mathbb{R}^n$, a fixed $\varepsilon > 0$, a positive number s, and admissible exponents $p, q \in C(\overline{\Omega})$ which satisfy the inequalities

$$n + \delta < p < q < p + \varepsilon$$

in Ω, for some $\delta > 0$, then with

$$s' := \varepsilon\,|\Omega| + \varepsilon^{-\varepsilon}s$$

there exists a positive constant $K = K(\Omega, q, s)$ such that

$$\frac{1}{s'}\lambda_{p,s'} \leq K^\varepsilon \frac{1}{s}\lambda_{q,s}.$$

Proof.   Let C be the norm of the Sobolev embedding

$$E : W_q^1(\Omega) \to L_\infty(\Omega),$$

(see Theorem 2.13) and denote the best constant in the q-Poincaré inequality by C', that is,

$$C' = \sup \frac{\|u\|_q}{\|\nabla u\|_q}$$

where the supremum is taken over all $u \in \overset{\circ}{W}_q^1(\Omega)\backslash\{0\}$.

Let $u_{q,s} \in \overset{\circ}{W}_q^1(\Omega)$ be the minimal function in V_s obtained by way of Lemma 4.3. From Lemma 4.6,

$$\rho_p\left(\frac{|\nabla u_{q,s}|}{q^{1/q}}\right) \leq \varepsilon^{-\varepsilon}s + \varepsilon|\Omega| := s'. \qquad (4.23)$$

On the other hand, on account of Lemma 4.5 and inequality (4.23),

$$
S_q(s) = \int_\Omega \frac{|u_{q,s}|^q}{q}\, dx \le \left(\left\| \frac{u_{q,s}}{q^{\frac{1}{q}}} \right\|_\infty + 1 \right)^\varepsilon \int_\Omega \frac{|u_{q,s}|^p}{q^{\frac{p}{q}}}\, dx
$$

$$
\le \left(C q_-^{-\frac{1}{q^+}} \|u_{q,s}\|_{1,q} + 1 \right)^\varepsilon S_p(s')
$$

$$
\le \left(C(C'+1) q_-^{-\frac{1}{q^+}} \|\nabla u_{q,s}\|_q + 1 \right)^\varepsilon S_p(s'). \quad (4.24)
$$

A straightforward application of Proposition 1.5 in conjunction with (4.24) yields

$$
S_q(s)
$$

$$
\le \left(C(C'+1) q_-^{-\frac{1}{q^+}} \max\left\{ \left(\int_\Omega |\nabla u_{q,s}|^q \right)^{\frac{1}{q^+}}, \left(\int_\Omega |\nabla u_{q,s}|^q \right)^{\frac{1}{q^-}} \right\} + 1 \right)^\varepsilon
$$

$$
\times S_p(s')
$$

$$
\le \left(C(C'+1) q_-^{-\frac{1}{q^+}} \max\left\{ (q_+ s)^{\frac{1}{q^+}}, (q_+ s)^{\frac{1}{q^-}} \right\} + 1 \right)^\varepsilon S_p(s'). \quad (4.25)
$$

Next,

$$
\lambda_{q,s} = \frac{\int_\Omega |\nabla u_{q,s}|^q\, dx}{\int_\Omega |u_{q,s}|^q\, dx} \ge \frac{q_- s}{q^+ S_q(s)} \ge \frac{\frac{q_-}{q^+} K^{-\varepsilon} s}{S_p(s')}
$$

$$
= \frac{q_-}{s'^+} K^{-\varepsilon} s \lambda_{p,s'}.
$$

Hence

$$
\lambda_{q,s} \ge \frac{q_-}{q^+} K^{-\varepsilon} \frac{s}{s'} \lambda_{p,s'},
$$

which quickly leads to the claim. $\qquad\square$

4.5 Convergence Properties of the Eigenfunctions

In this section we refine our study of the stability of the modular eigenvalues introduced in the previous section relative to the integrability exponent, in the sense described below (see for more details [58]). Consider a bounded Lipschitz domain $\Omega \subset \mathbb{R}^n$ and $\phi \in C(\overline{\Omega})$, subject to the condition

$$
1 < \inf_\Omega \phi = \phi_- \le \sup_\Omega \phi = \phi_+ < \infty.
$$

We stick to the convention of endowing the Sobolev space $\overset{\circ}{W}{}^1_\phi(\Omega)$ with the norm

$$u \longmapsto \|\,|\nabla u|\,\|_\phi.$$

Theorem 4.5 guarantees that for each such exponent ϕ there exists at least a maximal function u_ϕ for the constrained problem

$$\int_\Omega \frac{|u_\phi|^\phi}{\phi} = \max \int_\Omega \frac{|u|^\phi}{\phi} dx, \qquad (4.26)$$

(where the maximum is taken over all u such that $\int_\Omega \frac{|\nabla u|^\phi}{\phi} \leq 1$) which, in addition, fulfills the condition

$$\int_\Omega \frac{|\nabla u_\phi|^\phi}{\phi} = 1.$$

Consider two functions p and q in $C(\overline{\Omega})$, a non-decreasing sequence

$$(p_i) \subset C(\overline{\Omega})$$

such that

$$p_i(x) \to q(x)$$

uniformly in Ω and assume that

$$n < \inf_\Omega g = q_- \leq q_+ < \infty.$$

For each natural number i, B_i stands for the unit ball in $\overset{\circ}{W}{}^1_{p_i}(\Omega)$ furnished with the norm

$$\|u\|_{1,p_i} = \|\nabla u\|_{p_i}.$$

In what follows, we fix a sequence of maximal functions

$$(u_{p_i}) \subset \overset{\circ}{W}{}^1_{p_i}(\Omega)$$

satisfying (4.26) and

$$\int_\Omega \frac{|\nabla u_{p_i}|^{p_i}}{p_i} = 1$$

for each $i \in \mathbb{N}$. Clearly

$$\int_\Omega |\nabla u_{p_i}|^{p_i} \leq p_i^+ \int_\Omega \frac{|\nabla u_{p_i}|^{p_i}}{p_i} dx \leq q^+.$$

In view of the inequality in Lemma 4.6 it follows that

$$\int_\Omega |\nabla u_{p_i}|^p \leq \|p_i - p\|_\infty^{-\|p_i - p\|_\infty} \int_\Omega |\nabla u_{p_i}|^{p_i} + \|p_i - p\|_\infty |\Omega|$$
$$\leq \|p - q\|_\infty^{-\|p - q\|_\infty} q^+ + \|p - q\| |\Omega|.$$

Owing to Proposition 1.5, the sequence (u_{p_i}) is bounded in $\overset{\circ}{W}{}^1_p(\Omega)$; standard functional-analytic arguments imply that there exist $u \in \overset{\circ}{W}{}^1_p(\Omega)$ and a weakly-convergent subsequence of (u_{p_i}), still denoted by (u_{p_i}), such that

$$\lim_{i \to \infty} u_{p_i} = u$$

in $L_p(\Omega)$ and $u_{p_i} \to u$ pointwise a.e. in Ω.
Set

$$f(\sigma) = \sigma|\Omega| + \sigma^{-\sigma}(1 + \sigma).$$

It is easy to verify that f is increasing on the interval $(0, e^{-1})$ and that

$$\lim_{\sigma \to 0^+} f(\sigma) = 1;$$

hence, for arbitrary $\delta > 0$ there exists $\eta \in (0, 1)$ such that $0 < \sigma \leq \eta$ guarantees $f(\sigma) < 1 + \delta$. By assumption, $\|p_i - p_j\|_\infty \to 0$ as $i, j \to \infty$: select $k_0 \in \mathbb{N}$ such that $i, k \geq k_0$ implies $\|p_i - p_k\|_\infty < \eta$; in particular, for $i \geq k \geq k_0$ and each $x \in \Omega$, we have

$$1 < \frac{p_i(x)}{p_k(x)} < 1 + \eta. \tag{4.27}$$

Along the same lines as in the proof of Lemma 4.6, we have, for $k \leq i$:

$$\int_\Omega \frac{|\nabla u_{p_i}|^{p_k}}{p_k} \, dx =$$

$$\int_{\{x:|\nabla u_{p_i}(x)|<\eta\}} \frac{|\nabla u_{p_i}|^{p_k}}{p_k} \, dx$$

$$+ \int_{\{x:\eta \leq |\nabla u_{p_i}(x)|<1\}} |\nabla u_{p_i}|^{p_k - p_i} \frac{p_i}{p_k} \frac{|\nabla u_{p_i}|^{p_i}}{p_i} \, dx$$

$$+ \int_{\{x:|\nabla u_{p_i}(x)|>1\}} \frac{p_i}{p_k} \frac{|\nabla u_{p_i}|^{p_i}}{p_i} \, dx$$

$$\leq \eta|\Omega| + \eta^{-\eta} \int_{\{x:\eta \leq |\nabla u_{p_i}(x)|<1\}} \frac{p_i}{p_k} \frac{|\nabla u_{p_i}|^{p_i}}{p_i} \, dx$$

$$+ \int_{\{x:|\nabla u_{p_i}(x)|>1\}} \frac{p_i}{p_k} \frac{|\nabla u_{p_i}|^{p_i}}{p_i} \, dx$$

$$\leq \eta|\Omega| + \eta^{-\eta} \int_\Omega \frac{p_i}{p_k} \frac{|\nabla u_{p_i}|^{p_i}}{p_i} \, dx$$

$$\leq \eta|\Omega| + \eta^{-\eta} (1 + \eta) \int_\Omega \frac{|\nabla u_{p_i}|^{p_i}}{p_i} \, dx$$

$$< 1 + \delta. \tag{4.28}$$

It is clear from (4.28) that the sequence

$$(u_{p_i})_{i \geq k}$$

is bounded in $\overset{\circ}{W}{}_{p_k}^1(\Omega)$ for each fixed $k \geq k_0$. Hence there is a subsequence that converges weakly to $v \in \overset{\circ}{W}{}_{p_k}^1(\Omega)$; thus a fortiori one must have

$$v = u.$$

It therefore follows that for each $k \geq k_0$, it holds that

$$u \in \overset{\circ}{W}{}_{p_k}^1(\Omega)$$

and that there is a subsequence of (u_{p_i}) such that

$$u_{p_i} \rightharpoonup u \text{ in } \overset{\circ}{W}{}_{p_k}^1(\Omega)$$

and

$$u_{p_i} \longrightarrow u \text{ in } L_{p_k}(\Omega).$$

From the arbitrariness of k it follows that

$$u \in \bigcap_{k=1}^\infty \overset{\circ}{W}{}_{p_k}^1(\Omega).$$

Lemma 4.7. *The function* $u = \lim_{i \to \infty} u_{p_i}$ *defined in the above paragraph belongs to* $\overset{\circ}{W}{}^1_q(\Omega)$. *Moreover,*

$$\int_\Omega \frac{|\nabla u|^q}{q}\, dx \leq 1.$$

Proof. Let $\delta > 0$ be arbitrary and k_0 be as in the paragraph preceding inequality (4.27). The assumption on the sequence $(p_i)_i$ and Fatou's lemma yield

$$\int_\Omega \frac{|\nabla u|^q}{q} dx \leq \liminf_{k \geq k_0} \int_\Omega \frac{|\nabla u|^{p_k}}{p_k}\, dx.$$

On the other hand, since the functional

$$F_k : \overset{\circ}{W}{}^1_{p_k}(\Omega) \to [0, \infty)$$

given by

$$F_k(w) = \int_\Omega \frac{|\nabla w|^{p_k}}{p_k}\, dx$$

is weakly lower-semicontinuous (see Proposizione 4.2.2 in [16]), it follows easily that

$$F_k(u) \leq \liminf_{i \to \infty} F_k(u_i).$$

From the inequality (4.28) we have

$$\liminf_{i \geq k_0} \int_\Omega \frac{|\nabla u_{p_i}|^{p_k}}{p_k} \leq 1 + \delta;$$

the claims follows directly from the arbitrariness of δ. $\qquad\square$

Theorem 4.8. *Let the sequence* (u_{p_i}) *be as in Lemma 4.7. Then the limit function*

$$u = \lim_{i \to \infty} u_{p_i}$$

obtained therein is maximal in the following sense:

$$\int_\Omega \frac{|u|^q}{q} dx = \sup \int_\Omega \frac{|v|^q}{q} dx,$$

where the supremum is taken over all v *for which* $\int_\Omega |\nabla v|^q / q\, dx \leq 1$.

Proof. Fix $v \in \overset{\circ}{W}{}_q^1(\Omega)$ with $\int_\Omega |\nabla v|^q / q \, dx = 1$ and a sequence

$$(v_j) \subset C_0^\infty(\Omega)$$

converging to v in $\overset{\circ}{W}{}_q^1(\Omega)$ and chosen in such a way that

$$v_j \to v \ , \ \nabla v_j \to \nabla v$$

pointwise a.e. in Ω. Let $\delta > 0$ be arbitrary: pick $\eta > 0$ and $\theta > 0$ with

$$(1 + \eta)(1 + 2\theta) < 1 + \delta.$$

Lebesgue's dominated convergence theorem implies that

$$\lim_{k \to \infty} \int_\Omega \frac{|\nabla v|^{p_k}}{p_k^{\frac{q}{p_k}}} \, dx = \int_\Omega \frac{|\nabla v|^q}{q} \, dx = 1;$$

it is thus clear that $k \in \mathbb{N}$ can be chosen so large that it simultaneously satisfies

$$\|p_k - q\|_\infty^{-\|p_k - q\|_\infty} + \|p_k - q\|_\infty |\Omega| < 1 + \eta \tag{4.29}$$

and

$$\int_\Omega \frac{|\nabla v|^{p_k}}{p_k^{\frac{q}{p_k}}} \, dx < (1 + \theta)^{p-}.$$

The last inequality coupled with Proposition 1.5 gives

$$\left\| \frac{\nabla v}{p_k^{\frac{1}{p_k}}} \right\|_q \leq (1 + \delta)^{\frac{p-}{q-}} \leq 1 + \theta. \tag{4.30}$$

Next, select j large enough to satisfy

$$\|\nabla(v_j - v)\|_q < \theta. \tag{4.31}$$

Lemma 1.4 gives

$$\left\| \frac{\nabla v_j}{p_k^{\frac{1}{p_k}}} \right\|_{p_k} \leq \left(\|p_k - q\|_\infty^{-\|p_k - q\|_\infty} + \|p_k - q\|_\infty |\Omega| \right)$$

$$\times \left(\left\| \frac{\nabla(v_j - v)}{p_k^{\frac{1}{p_k}}} \right\|_q + \left\| \frac{\nabla v}{p_k^{\frac{1}{p_k}}} \right\|_q \right),$$

whence by virtue of (4.29), (4.30) and (4.31) we have

$$\left\| \frac{\nabla v_j}{p_k^{\frac{1}{p_k}}} \right\|_{p_k} \leq (1 + \eta)(1 + 2\theta) < 1 + \delta,$$

that is,

$$\left\| (1+\delta)^{-1} \frac{\nabla v_j}{p_k^{1/p_k}} \right\|_{p_k} \leq 1.$$

Hence

$$\int_\Omega (1+\delta)^{-p_k} \frac{|\nabla v_j|^{p_k}}{p_k} \, dx \leq 1.$$

The maximality of u_{p_k} yields

$$\int_\Omega \left| \frac{v_j}{1+\delta} \right|^{p_k} \frac{1}{p_k} \, dx \leq \int_\Omega \frac{|u_{p_k}|^{p_k}}{p_k} \, dx,$$

from which it is immediate that

$$\int_\Omega \frac{|v_j|^{p_k}}{p_k} \, dx \leq (1+\delta)^{p_k^+} \int_\Omega \frac{|u_{p_k}|^{p_k}}{p_k} \, dx. \qquad (4.32)$$

The assumption $\inf_\Omega q(x) > n$ implies that $\|u_{p_k}\|_\infty \leq C$ uniformly in k; since (u_{p_k}) converges pointwise almost everywhere to u in Ω, Lebesgue's dominated convergence theorem gives

$$\int_\Omega \frac{|u_{p_k}|^{p_k}}{p_k} \, dx \to \int_\Omega \frac{|u|^q}{q} \, dx.$$

Letting $k \to \infty$ in (4.32) for fixed j we see that

$$\int_\Omega \frac{|v_j|^q}{q} \, dx \leq (1+\delta)^{q^+} \int_\Omega \frac{|u|^q}{q} \, dx.$$

In conclusion

$$\int_\Omega \frac{|v|^q}{q} \, dx \leq (1+\delta)^{q^+} \int_\Omega \frac{|u|^q}{q} \, dx,$$

for arbitrary $\delta > 0$ and therefore u is maximal. $\qquad \square$

Remark 4.1. It follows from Theorem 4.5 that the maximal character of u implies

$$\int_\Omega \frac{|\nabla u|^q}{q} \, dx = 1. \qquad (4.33)$$

Theorem 4.9. *Let (p_i) be a non-decreasing sequence of functions in $C(\overline{\Omega})$ with*

$$\lim_{i \to \infty} p_i = q \in C(\overline{\Omega}).$$

Assume furthermore that $\inf_\Omega q = q_- > n$. *For each* $i \in \mathbb{N}$ *let* $u_i \in \overset{\circ}{W}{}^1_{p_i}(\Omega)$ *be an eigenfunction corresponding to the least eigenvalue* $\lambda_{p_i,1,\Omega}$ *of the* p_i-*Laplacian (see Definition 4.1). Then there exists a subsequence, still denoted by* (u_i), *that converges weakly to* $u \in \overset{\circ}{W}{}^1_q(\Omega)$ *where*

$$\int_\Omega \frac{|u|^q}{q}\, dx = \sup \int_\Omega \frac{|v|^q}{q}\, dx,$$

the supremum being taken over all v *for which*

$$\int_\Omega \frac{|\nabla v|^q}{q}\, dx \le 1.$$

Moreover

$$\int_\Omega \frac{|\nabla u|^q}{q}\, dx = 1.$$

and the limit

$$\lim_{i \longrightarrow \infty} \lambda_{p_i,1} = \lim_{i \to \infty} \frac{\int_\Omega |\nabla u_i|^{p_i}\, dx}{\int_\Omega |u_i|^{p_i}\, dx} = L < \infty$$

exists. The limit function u *satisfies*

$$\frac{q_-}{q_+} L \le \frac{\int_\Omega |\nabla u|^q\, dx}{\int_\Omega |u|^q\, dx} \le L, \tag{4.34}$$

and

$$\left(\frac{q_-}{q_+}\right)^2 L \le \lambda_{q,1} \le L.$$

Proof.   In the course of the proof we retain the terminology of Theorem 4.8. On account of Theorem 4.8 and because of the maximality of u we have

$$\int_\Omega \frac{|u_i|^{p_i}}{p_i} \to \int_\Omega \frac{|u|^q}{q},$$

from which it follows that

$$\int_\Omega |u_i|^{p_i} \to \int_\Omega |u|^q.$$

From the above considerations, the conditions on the sequence (p_i) and the bounds

$$\frac{p_{i-}}{\int_\Omega |u_i|^{p_i}} \le \frac{\int_\Omega |\nabla u_i|^{p_i}}{\int_\Omega |u_i|^{p_i}} \le \frac{p_{i+}}{\int_\Omega |u_i|^{p_i}},$$

it is clear that the numerical sequence

$$\left(\frac{\int_\Omega |\nabla u_i|^{p_i}}{\int_\Omega |u_i|^{p_i}} \right) \tag{4.35}$$

is bounded; hence it is either finite or a subsequence of the sequence (u_i) can be extracted so that the sequence of quotients (4.35) is convergent, say to $L \in \mathbb{R}$. Then

$$\frac{\int_\Omega |\nabla u|^q}{\int_\Omega |u|^q} \leq \frac{q^+ \int_\Omega |\nabla v|^q}{q^- \int_\Omega |v|^q}.$$

Next, from the equality

$$\lim_{\eta \to 0} \eta^{-\eta} = 1,$$

it follows that given $\delta > 0$, there exists $\mu > 0$ such that if $0 < \eta < \mu$, then

$$1 \leq \eta^{-\eta} < 1 + \delta.$$

Since $p_i \to q$ pointwise uniformly in Ω, there is a natural number N such that $i, j \geq N$ implies

$$\|p_i - p_j\|_\infty < \mu' = \min \left\{ \mu, \delta|\Omega|^{-1} \int_\Omega |u|^q \, dx \right\}.$$

Fatou's lemma guarantees that

$$\int_\Omega |\nabla u|^q \, dx \leq \liminf_{k \to \infty} \int_\Omega |\nabla u|^{p_k} \, dx;$$

on the other hand the weak lower-semicontinuity of the functional on $\overset{\circ}{W}{}^1_{p_k}(\Omega)$ introduced in Corollary 4.1 yields

$$\int_\Omega |\nabla u|^{p_k} \, dx \leq \liminf_{i \geq k} \int_\Omega |\nabla u_i|^{p_k} \, dx.$$

Now, Lemma 4.6 together with the above considerations imply that for fixed $k \geq N$ and all $i \geq k$,

$$\int_\Omega |\nabla u_i|^{p_k} \, dx \leq (1+\delta) \int_\Omega |\nabla u_i|^{p_i} \, dx + \delta \int_\Omega |u|^q \, dx.$$

In other words, for $i \geq k \geq N$ we have the inequality:

$$\frac{\int_\Omega |\nabla u_i|^{p_k}}{\int_\Omega |u_i|^{p_i}} \leq (1+\delta) \frac{\int_\Omega |\nabla u_i|^{p_i}}{\int_\Omega |u_i|^{p_i}} + \delta \frac{\int_\Omega |u|^q}{\int_\Omega |u_i|^{p_i}}.$$

In view of (4.5), we have thus proved that for i and k as above,

$$\frac{\liminf_{i \geq k} \int_\Omega |\nabla u_i|^{p_k}}{\int_\Omega |u|^q} = \liminf_{i \geq k} \frac{\int_\Omega |\nabla u_i|^{p_k}}{\int_\Omega |u_i|^{p_i}} \leq (1+\delta)L + \delta.$$

Hence for arbitrary $\delta > 0$ there exists N such that for all natural numbers k with $k \geq N$,

$$\frac{\int_\Omega |\nabla u|^{p_k}\,dx}{\int_\Omega |u|^q\,dx} \leq (1+\delta)L + \delta.$$

Invoking Fatou's lemma, it is now easy to see that

$$\frac{\int_\Omega |\nabla u|^q\,dx}{\int_\Omega |u|^q\,dx} \leq \liminf_{k\to\infty} \frac{\int_\Omega |\nabla u|^{p_k}\,dx}{\int_\Omega |u|^{q(x)}\,dx} \leq (1+\delta)L + \delta$$

and so

$$\frac{\int_\Omega |\nabla u|^q\,dx}{\int_\Omega |u|^q\,dx} \leq L. \tag{4.36}$$

On the other hand, (4.33), (4.5) and the equalities

$$\int_\Omega \frac{|\nabla u_i|^{p_i}}{p_i} = 1 \quad (i \in \mathbb{N})$$

guarantee that given $\delta > 0$,

$$\frac{\int_\Omega |\nabla u|^q}{\int_\Omega |u|^q} \geq \frac{q_-\int_\Omega \frac{|\nabla u|^q}{q}}{\int_\Omega |u|^q} = \frac{q_-}{\int_\Omega |u|^q} \geq \frac{q_-\int_\Omega \frac{|\nabla u_i|^{p_i}}{p_i}}{\int_\Omega |u_i|^{p_i} + \delta} \geq \frac{q_-/q_+\int_\Omega |\nabla u_i|^{p_i}}{\int_\Omega |u_i|^{p_i} + \delta}$$

for sufficiently large i. Coupled with (4.36) this implies (4.34).

Next, we fix an arbitrary function

$$v \in V_{q,1} \subseteq \overset{\circ}{W}{}^1_{p_k}(\Omega).$$

The maximality of v gives

$$\int_\Omega \frac{|u_i|^{p_i}}{p_i} \longrightarrow \int_\Omega \frac{|v|^q}{q}.$$

Thus for $\delta > 0$ and i large enough,

$$\frac{\int_\Omega |\nabla v|^q}{\int_\Omega |v|^q} \geq q_-/q_+ \frac{\int_\Omega |\nabla v|^q/q}{\int_\Omega |v|^q/q}$$

$$= q_-/q_+ \frac{\int_\Omega |\nabla v_i|^q/q}{\int_\Omega |v_i|^q/q}$$

$$\geq \left(\frac{q_-p_{i-}}{q_+p_{i+}}\right) \frac{\int_\Omega |\nabla u_i|^{p_i}}{\int_\Omega |u_i|^{p_i} + \delta}.$$

Letting $i \longrightarrow \infty$ in the above inequalities we obtain

$$\frac{\int_\Omega |\nabla v|^q}{\int_\Omega |v|^q} \geq (q_-/q_+)^2 L.$$

Finally, the arbitrariness of v yields

$$\lambda_{q,1} \geq (q_-/q_+)^2 L.$$

The latter inequality completes the proof of the Theorem.   □

If $\inf_{x \in \Omega} q(x) \geq 2$, a stronger convergence result holds, as discussed in the following theorem.

Theorem 4.10. *For $q_- \geq 2$, u, (p_i) and (u_i) as in Theorem 4.9, one has*

$$\lim_{j \to \infty} \int_\Omega |\nabla u - \nabla u_j|^{p_j} \, dx = 0.$$

In particular:

$$L = \frac{\int_\Omega |\nabla u|^q}{\int_\Omega |u|^q}.$$

Proof. For each fixed natural number k, the weak lower-semicontinuity of the functional

$$G_k : \overset{\circ}{W}{}^1_{p_k}(\Omega) \to [0, \infty)$$

$$G_k(v) = \int_\Omega \frac{|\nabla v|^{p_k}}{p_k} dx$$

implies that

$$\int_\Omega \frac{|\nabla u|^{p_k}}{p_k} dx \leq \liminf_{j \to \infty} \int_\Omega \left| \frac{\nabla(u + u_j)}{2} \right|^{p_k} \frac{1}{p_k} \, dx.$$

The embedding estimate (4.22) yields, for $j \geq k$ and $\varepsilon = \|p_j - p_k\|_\infty$,

$$\liminf_{j \to \infty} \int_\Omega \left| \frac{\nabla(u + u_j)}{2} \right|^{p_k} \frac{1}{p_k}$$

$$\leq \varepsilon^{-\varepsilon} \liminf_{j \to \infty} \int_\Omega \left| \frac{\nabla(u + u_j)}{2} \right|^{p_j} \left(\frac{1}{p_k^{p_j/p_k}} - \frac{1}{p_j} \right)$$

$$+ \varepsilon^{-\varepsilon} \liminf_{j \to \infty} \int_\Omega \left| \frac{\nabla(u + u_j)}{2} \right|^{p_j} \frac{1}{p_j} + \varepsilon |\Omega|.$$

We observe that

$$\int_\Omega \left| \frac{\nabla(u + u_j)}{2} \right|^{p_j} \frac{1}{p_j} \leq \frac{1}{2} \int_\Omega \frac{|\nabla u|^{p_j}}{p_j} + \frac{1}{2} \int_\Omega \frac{|\nabla u_j|^{p_j}}{p_j} \tag{4.37}$$

$$= \frac{1}{2} \int_\Omega \frac{|\nabla u|^{p_j}}{p_j} + \frac{1}{2}.$$

A straightforward application of Lebesgue's dominated convergence theorem reveals that the right-hand side in (4.37) is bounded as $j \to \infty$. On the other hand,

$$\varepsilon^{-\varepsilon} \longrightarrow 1 \quad \text{as} \quad \varepsilon \longrightarrow 0.$$

Consequently:

$$\varepsilon^{-\varepsilon} \liminf_{j \to \infty} \int_\Omega \left| \frac{\nabla(u + u_j)}{2} \right|^{p_j} \left(\frac{1}{p_k^{p_j/p_k}} - \frac{1}{p_j} \right) \to 0 \text{ as } k, j \to \infty.$$

Fix $\delta > 0$ and choose $J > 0$ large enough so that $j \geq k \geq J$ implies

$$\left| \varepsilon^{-\varepsilon} \int_\Omega \left| \frac{\nabla(u + u_j)}{2} \right|^{p_j} \left(\frac{1}{p_k^{p_j/p_k}} - \frac{1}{p_j} \right) \right| < \delta,$$

$\|p_k - p_j\|_\infty |\Omega| < \delta$ and

$$\|p_k - p_j\|_\infty^{-\|p_k - p_j\|_\infty} < 1 + \delta.$$

Set

$$T = \liminf_{j \to \infty} \int_\Omega \left| \frac{\nabla(u + u_j)}{2} \right|^{p_j} \frac{1}{p_j}$$

and pick I large enough for $i \geq I$ to guarantee that

$$T - \delta < \inf_{j \geq i} \int_\Omega \left| \frac{\nabla(u + u_j)}{2} \right| \frac{1}{p_j}.$$

Then, for $j \geq \max\{k, I, J\}$,

$$\int_\Omega \frac{|\nabla u|^{p_k}}{p_k} dx$$

$$\leq (1 + \delta) \left(\int_\Omega \left| \frac{\nabla(u + u_j)}{2} \right|^{p_j} \frac{1}{p_j} + \delta \right) + 2\delta$$

$$\leq (1 + \delta) \left(\delta + \frac{1}{2} \left(\int_\Omega \frac{|\nabla u|^{p_j}}{p_j} + \int_\Omega \frac{|\nabla u_j|^{p_j}}{p_j} \right) \right) + 2\delta.$$

Thus, since

$$\int_\Omega \frac{|\nabla u|^{p_i}}{p_i} \longrightarrow \int_\Omega \frac{|\nabla u|^q}{q} = 1 \text{ as } i \longrightarrow \infty,$$

$$\lim_{j \to \infty} \int_\Omega \left| \frac{\nabla(u + u_j)}{2} \right|^{p_j} \frac{1}{p_j} = 1. \tag{4.38}$$

Under the hypothesis $q_- \geq 2$ and from the uniform convergence of the sequence (p_i), it can be assumed that for $j \geq J$ (J as above),

$$\inf_\Omega p_j \geq 2.$$

We now invoke the inequalities

$$\left| \frac{a + b}{2} \right|^t + \left| \frac{a - b}{2} \right|^t \leq \frac{1}{2} \left(|a|^t + |b|^t \right)$$

valid for any real number $t : t \geq 2$ and any complex numbers a and b (see [4], Lemma 2.27). In particular:

$$\int_{\Omega} \left| \frac{\nabla(u + u_j)}{2p_j^{1/p_j}} \right|^{p_j} dx + \int_{\Omega} \left| \frac{\nabla(u - u_j)}{2p_j^{1/p_j}} \right|^{p_j} dx$$

$$\leq \frac{1}{2} \left(\int_{\Omega} \left| \frac{\nabla u}{p_j^{1/p_j}} \right|^{p_j} + \int_{\Omega} \left| \frac{\nabla u_j}{p_j^{1/p_j}} \right|^{p_j} \right).$$

Letting $j \to \infty$ and taking (4.38) into account, it follows easily that;

$$\lim_{j \to \infty} \int_{\Omega} |\nabla u - \nabla u_j|^{p_j} = 0. \tag{4.39}$$

This completes the proof of Theorem 4.10. As for the Corollary, we start by observing that with the aid of Lebesgue's theorem it is easy to see that

$$\int_{\Omega} |\nabla u|^{p_j} \longrightarrow \int_{\Omega} |\nabla u|^q.$$

Hence, bringing the convexity of the modular into play and taking any real number $t \in (0, 1)$ we have

$$\int_{\Omega} |\nabla u_j|^{p_j} - \int_{\Omega} |\nabla u|^{p_j}$$

$$\leq t^{1-q+} \int_{\Omega} |\nabla(u_j - u)|^{p_j}$$

$$+ \left((1-t)^{1-q+} - 1 \right) \left(\int_{\Omega} |\nabla u|^{p_j} - \int_{\Omega} |\nabla u|^q \right)$$

$$+ \left((1-t)^{1-q+} - 1 \right) \int_{\Omega} |\nabla u|^q.$$

Likewise

$$\int_{\Omega} |\nabla u|^{p_j} - \int_{\Omega} |\nabla u_j|^{p_j} \leq t^{1-q+} \int_{\Omega} |\nabla(u_j - u)|^{p_j} + \left((1-t)^{1-q+} - 1 \right) \int_{\Omega} |\nabla u_j|^{p_j}.$$

Since

$$\int_{\Omega} |\nabla u_j|^{p_j} \leq q_+,$$

it follows from the above two estimates that

$$\int_{\Omega} |\nabla u_j|^{p_j} - \int_{\Omega} |\nabla u|^{p_j} \longrightarrow 0 \text{ as } j \longrightarrow \infty.$$

Hence via equality (4.39) it is apparent that

$$\int_{\Omega} |\nabla u_j|^{p_j} \longrightarrow \int_{\Omega} |\nabla u|^q \text{ as } j \longrightarrow \infty.$$

It is therefore clear that

$$L \leq \frac{\int_\Omega |\nabla u|^q}{\int_\Omega |u|^q} = \lim_{j \longrightarrow \infty} \frac{\int_\Omega |\nabla u_j|^{p_j}}{\int_\Omega |u_j|^{p_j}} = \lim_{j \longrightarrow \infty} \lambda_{p_j,1};$$

this inequality coupled with (4.36) yields the claim. $\qquad\square$

The condition $p_j > n$ cannot be removed: Lindqvist [62] constructed a bounded domain for which the convergence statement of Theorem 4.10 fails even for constant $p \leq n$.

We single out a particular case of the preceding Theorem.

Theorem 4.11. *Let $n = 1$, $\Omega = (a, b)$, $a < b$ and suppose that*

$$(p_j) \subset C([a, b])$$

is a non-decreasing sequence of functions, uniformly convergent to

$$q = \sup_j p_j,$$

with

$$1 < p_{1-} \ \text{ and } \ q_+ < \infty.$$

For each $j \in \mathbb{N}$ let λ_j and $u_j \in \overset{\circ}{W}{}^1_{p_j}(\Omega)$ be respectively the first eigenvalue and the first eigenfunction of the p_j Laplacian, namely:

$$\left(|u'|^{p_j-2}u'\right)' = \lambda_j |u|^{p_j-2}u;$$

likewise, denote the first eigenvalue of the q-Laplacian by $\lambda \in \mathbb{R}$. Then there exists a subsequence (u_k) of (u_j) and a first eigenfunction

$$u \in \overset{\circ}{W}{}^1_q(\Omega)$$

of the q-Laplacian such that

$$\lim_{k \to \infty} \lambda_k = \frac{\int_\Omega |u'|^q}{\int_\Omega |u|^q}.$$

Proof. For $q_- \geq 2$, the theorem is a direct consequence of Theorem 4.10; it remains to consider the case when $q_- < 2$. To this end we notice the elementary fact that there exist positive constants c, M, s and S such that for all s, M, t, r with

$$(t, r) \in (s, S) \times [-M, M],$$

the inequality

$$\frac{1}{2}\left(|r + 1|^t + |r - 1|^t\right) - |r|^t \geq c \tag{4.40}$$

holds. Fix $\delta > 0$ and set

$$A = \{x : (a,b) : |u_j{}'(x) - u{}'(x)| > \delta |u_j{}'(x) + u{}'(x)|\}.$$

Writing

$$r = \frac{u_j{}'(x) + u{}'(x)}{u_j{}'(x) - u{}'(x)} \quad , \quad t = p_j$$

in (4.40) one obtains:

$$\frac{1}{2}\left(|2u{}'|^{p_j} + |2u_j{}'|^{p_j}\right) - |u_j{}'(x) + u{}'(x)|^{p_j} \geq c|u_j{}'(x) - u{}'(x)|^{p_j}$$

or after multiplication of both sides by $1/(2^{p_j} p_j)$:

$$\frac{1}{2}\left(\frac{|u{}'|^{p_j}}{p_j} + \frac{|u_j{}'|^{p_j}}{p_j}\right) - \left|\frac{u_j{}' + u{}'}{2}\right|^{p_j}\frac{1}{p_j} \geq c\left|\frac{u_j{}' - u{}'}{2}\right|^{p_j}\frac{1}{p_j}. \qquad (4.41)$$

Integrating both sides of (4.41) over Ω it follows that

$$\frac{1}{2}\left(\int_\Omega \frac{|u{}'|^{p_j}}{p_j} + \int_\Omega \frac{|u_j{}'|^{p_j}}{p_j}\right) - \int_\Omega \left|\frac{u_j{}' + u{}'}{2}\right|^{p_j}\frac{1}{p_j} \qquad (4.42)$$

$$\geq c\int_A \left|\frac{u_j{}' - u{}'}{2}\right|^{p_j}\frac{1}{p_j}.$$

As was shown in the proof of Theorem 4.10,

$$\int_\Omega \left|\frac{u_j{}' + u{}'}{2}\right|^{p_j}\frac{1}{p_j} \to 1 \text{ as } j \to \infty,$$

whereas

$$\int_\Omega \frac{|u|^{p_j}}{p_j} \longrightarrow 1 \text{ as } j \to \infty.$$

Letting $j \longrightarrow \infty$ in (4.42), and noting that $1 < p_j < q_+ < \infty$ in Ω, we see that

$$\int_A \left|\frac{u_j{}' - u{}'}{2}\right|^{p_j}\frac{1}{p_j} \longrightarrow 0 \text{ as } j \longrightarrow \infty. \qquad (4.43)$$

On the other hand,

$$\lim_{j\to\infty}\int_{\Omega\setminus A}\left|\frac{u{}' - u_j{}'}{2}\right|^{p_j}\frac{1}{p_j} \leq \lim_{j\to\infty}\delta^{p_j}\int_\Omega\left|\frac{u{}' + u_j{}'}{2}\right|^{p_j}\frac{1}{p_j} \leq \delta^{q_-};$$

the arbitrariness of δ and (4.43) then yields

$$\lim_{j\to\infty}\int_\Omega |u{}' - u_j{}'|^{p_j} = 0.$$

The statement concerning the convergence of the eigenvalues follows in the same way as the corresponding assertion in Theorem 4.10. $\qquad\qquad \square$

We now turn our attention to the case of a non-decreasing sequence (p_j) in Ω.

Lemma 4.8. *Consider a non-increasing sequence $(p_j) \subset C(\overline{\Omega})$ uniformly convergent in Ω to its infimum p; assume $1 < p_- \le p_+ < \infty$ and that for some constant $\theta > 0$ and a function $r \in C(\overline{\Omega})$,*

$$p(x) + \theta < r(x) \ \text{in } \Omega.$$

If a sequence $(\psi_j) \subset L_r(\Omega)$ converges to ψ in $L_r(\Omega)$, then

$$\limsup_j \int_\Omega \frac{|\psi_j|^{p_j}}{p_j} \le \int_\Omega \frac{|\psi|^p}{p}.$$

Proof.   In the light of Corollary 1.4, the sequence $\psi_j \longrightarrow \psi$ can be chosen to converge a.e in Ω. Fix the subindex $I \in \mathbb{N}$ so that the inequality

$$p_I < p + \theta < r$$

holds uniformly in Ω. According to Lemma 4.6, for any integer $j \ge I$ we have the estimate

$$
\int_\Omega \frac{|\psi_j|^{p_j}}{p_j} \le \|p_I - p_j\|_\infty^{-\|p_I - p_j\|_\infty} \int_\Omega |\psi_j|^{p_I} \left(\frac{1}{p_j^{p_I/p_j}} - \frac{1}{p_I^{p_I/p_j}} \right)
$$

$$
+ \|p_I - p_j\|_\infty^{-\|p_I - p_j\|_\infty} \int_\Omega \frac{|\psi_j|^{p_I}}{p_I^{p_I/p_j}} + \|p_I - p_j\|_\infty |\Omega|. \qquad (4.44)
$$

Since $p_I < r$ and (ψ_j) is convergent to ψ in $L_r(\Omega) \hookrightarrow L_{p_I}(\Omega)$, the pointwise estimate

$$|\psi_j|^{p_I} \le |\psi_j|^{p_I} \chi_{\{x: |\psi_j(x)| < 1\}} + |\psi_j|^r \chi_{\{x: |\psi_j(x)| \ge 1\}}$$

holds a.e. in Ω. Through a straightforward application of Lebesgue's dominated convergence theorem, the right-hand side of (4.44) can be seen to converge to

$$
\|p_I - p\|_\infty^{-\|p_I - p\|_\infty} \int_\Omega |\psi|^{p_I} \left(\frac{1}{p^{p_I/p}} - \frac{1}{p_I^{p_I/p}} \right) + \|p_I - p\|_\infty |\Omega|
$$

$$
+ \|p_I - p\|_\infty^{-\|p_I - p\|_\infty} \int_\Omega \frac{|\psi|^{p_I}}{p_I^{p_I/p}}.
$$

The obvious inequality

$$|\psi|^{p_I} \le |\psi|^{p_I} \chi_{\{x: |\psi(x)| < 1\}} + |\psi|^r \chi_{\{x: |\psi(x)| \ge 1\}}$$

in conjunction with Lebesgue's theorem yields that (4.44) tends to 0 as $I \to \infty$. This proves the Lemma.

$\square$

Theorem 4.12. *Let $\Omega \subset \mathbb{R}^n$ be a bounded Lipschitz domain. Consider a non-increasing sequence*

$$(p_j) \subset C(\overline{\Omega})$$

uniformly convergent in Ω to its infimum

$$p = \inf_j p_j;$$

assume $1 < p_- \leq p_+ < \infty$. Let

$$\lambda_{p,1}$$

be the first eigenvalue for the modular p-Laplacian introduced in Definition 4.1 and let u_j stand for the maximal function obtained in Lemma 4.3. For each $j \in \mathbb{N}$ let u_j be the maximal function given by Lemma 4.3. Then there exists

$$u \in V_{p,1}$$

and a subsequence of (u_j) (still denoted by (u_j)) such that, as $j \longrightarrow \infty$,

$$u_j \rightharpoonup u \ \ in \ \ \overset{\circ}{W}{}^1_p(\Omega)$$

and

$$u_j \to u \ \ in \ \ L_p(\Omega).$$

Furthermore, the limit

$$\lim_{i \longrightarrow \infty} \lambda_{p_i,1} = \lim_{i \to \infty} \frac{\int_\Omega |\nabla u_i|^{p_i}}{\int_\Omega |u_i|^{p_i}} = L < \infty$$

exists. The limit function u satisfies

$$\frac{p_-}{p_+} L \leq \frac{\int_\Omega |\nabla u|^p}{\int_\Omega |u|^p} \leq L. \tag{4.45}$$

Also

$$\lambda_{p,1} \leq \lim_{i \longrightarrow \infty} \lambda_{p_i,1} \leq \left(\frac{p_+}{p_-}\right)^2 \lambda_{p,1}. \tag{4.46}$$

Proof. By assumption, for each $j \in \mathbb{N}$, $u_j \in V_{p_j,1}$ (defined in (4.20)). In particular,

$$\int_\Omega \frac{|u_j|^{p_j}}{p_j} = \max_{\int_\Omega |\nabla v|^{p_j}/p_j \leq 1} \int_\Omega \frac{|v|^{p_j}}{p_j}.$$

According to Theorem 4.5 one must necessarily have:

$$\int_\Omega \frac{|\nabla u_j|^{p_j}}{p_j} = 1;$$

in particular, Lemma 4.6 implies that each u_j is subject to the bound

$$\int_\Omega |\nabla u_j|^p \leq (p_{1+})\|p_j - p\|_\infty^{-\|p_j-p\|_\infty} + \|p_j - p\|_\infty |\Omega|.$$

The assumptions on the sequence (p_j) and Proposition 1.5 imply that the sequence (u_j) is bounded in $\overset{\circ}{W}_p^1(\Omega)$.

Without loss of generality, the sequence (u_j) can then be assumed to be weakly convergent to $u \in \overset{\circ}{W}_p^1(\Omega)$ and Theorem 2.15 implies that in addition (u_j) can be taken to be strongly convergent to u in $L_p(\Omega)$. Corollary 1.4 implies, in turn, that

$$u_j \longrightarrow u \text{ as } j \longrightarrow \infty$$

pointwise almost everywhere in Ω.

For arbitrary $\delta > 0$, pick N large enough to guarantee that $j \geq N$ implies that

$$\|p_j - p\|_\infty^{-\|p_j-p\|_\infty} + \|p_j - p\|_\infty |\Omega| + < 1 + \delta$$

and that

$$\left| \frac{1}{p^{p_j/p}} - \frac{1}{p_j} \right| < \delta \text{ uniformly in } \Omega.$$

Then

$$\int_\Omega \frac{|\nabla u_j|^p}{p} \leq \|p_j - p\|_\infty^{-\|p_j-p\|_\infty} \left(\int_\Omega |\nabla u_j|^{p_j} \left(\frac{1}{p^{p_j/p}} - \frac{1}{p_j} \right) + \int_\Omega \frac{|\nabla u_j|^{p_j}}{p_j} \right)$$

$$+ \|p_j - p\|_\infty |\Omega|$$

$$\leq (1 + \delta)(p_{1+}\delta + 1).$$

The weak convergence

$$u_j \rightharpoonup u \text{ in } \overset{\circ}{W}_p^1(\Omega)$$

together with the weak lower-semicontinuity of the functional

$$F(w) = \int_\Omega \frac{|\nabla w|^p}{p}$$

yield

$$\int_\Omega \frac{|\nabla u|^p}{p} \leq (1 + \delta)(p_{1+}\delta + 1);$$

the arbitrariness of δ implies that

$$\int_\Omega \frac{|\nabla u|^p}{p} \leq 1.$$

Next, we show that u is maximal in the sense of Theorem 4.5.
Fix $\omega \in \overset{\circ}{W}{}^1_p(\Omega)$ such that

$$\int_\Omega \frac{|\nabla \omega|^p}{p} = 1;$$

pick a sequence of smooth functions $(\omega_k) \subset C_0^\infty(\Omega)$ converging to ω in $\overset{\circ}{W}{}^1_p(\Omega)$ and such that, as $k \to \infty$,

$$\nabla \omega_k \to \nabla \omega \text{ and } \omega_k \to \omega \text{ a.e in } \Omega.$$

For arbitrary $\delta > 0$ let k be large enough to satisfy:

$$1 - \delta/2 < \int_\Omega \frac{|\nabla \omega_k|^p}{p} < 1 + \delta/2. \tag{4.47}$$

Lebesgue's dominated convergence theorem yields, for each fixed k, that

$$\int_\Omega \frac{|\nabla \omega_k|^{p_j}}{p_j} \longrightarrow \int_\Omega \frac{|\nabla \omega_k|^p}{p} \text{ as } j \longrightarrow \infty.$$

It follows then from (4.47) and inequalities (4.11) and (4.12) that

$$\int_\Omega \frac{|\nabla \omega_k|^p}{p} \longrightarrow \int_\Omega \frac{|\nabla \omega|^p}{p} = 1 \text{ as } k \longrightarrow \infty.$$

Thus, for k fixed, j can be chosen so large that

$$\int_\Omega \frac{|\nabla \omega_k|^{p_j}}{p_j} \leq 1 + \delta.$$

The last inequality shows that for fixed $k \in \mathbb{N}$ there exists $J \in \mathbb{N}$ such that $j \geq J$ implies

$$\int_\Omega \frac{|\nabla \left(w_k/(1+\delta)^{1/p_j-}\right)|^{p_j}}{p_j} \leq 1,$$

whence the maximality of u_j gives for $j \geq J$,

$$\int_\Omega \frac{|w_k|^{p_j}}{(1+\delta)^{\frac{p_j}{p_j-}} p_j} \leq \int_\Omega \frac{|u_j|^{p_j}}{p_j}. \tag{4.48}$$

To facilitate the presentation, the proof will be split into the cases $n < p_-$, $p_- \leq n \leq p_+$, and $p_+ < n$. For $p_+ < n$, there exists a positive number $M \in \mathbb{N}$ sufficiently large for the inequality

$$\|p_M - p\|_\infty < \frac{1}{n-1}$$

to hold. Then, uniformly in Ω,

$$p_M \leq p + \frac{1}{n-1} < \frac{np}{n-p}.$$

Sobolev's embedding Theorem (Theorem 2.5) yields that for

$$r = \frac{np}{n-p} - \frac{1}{2(n-1)}$$

it holds that

$$\overset{\circ}{W}_p^1(\Omega) \hookrightarrow\hookrightarrow L_r(\Omega);$$

therefore no generality is lost by assuming that

$$(u_j) \subset L_r(\Omega)$$

and that (u_j) converges in $L_r(\Omega)$. Clearly

$$\lim_{j\to\infty} u_j = u \text{ in } L_r(\Omega).$$

Lemma 4.8 coupled with (4.48) yields

$$\int_\Omega \frac{|w_k|^p}{(1+\delta)^{\frac{p}{p_-}} p} = \limsup_{j \geq J} \int_\Omega \frac{|w_k|^{p_j}}{(1+\delta)^{\frac{p_j}{p_{j-}}} p_j} \leq \int_\Omega \frac{|u|^p}{p},$$

whence letting $k \to \infty$ it is apparent that

$$\int_\Omega \frac{|w|^p}{p} \leq (1+\delta)^{\frac{p}{p_-}} \int_\Omega \frac{|u|^p}{p}.$$

It is clear from the last inequality that u is maximal in the desired sense. We next proceed to prove the following stronger result:

$$\lim_{j\to\infty} \int_\Omega |u_j|^{p_j} = \int_\Omega |u|^p.$$

Indeed, for $p_j \leq r$, we have pointwise in Ω:

$$|u|^{p_j} \leq \chi_{\{x:|u(x)|>1\}} |u|^r + 1,$$

which via a direct application of Lebesgue's theorem leads to

$$\int_\Omega |u|^{p_j} \longrightarrow \int_\Omega |u|^p.$$

Since $(u_j)_{p_j \leq r}$ converges to u in $L_r(\Omega)$ the estimate

$$\|u - u_j\|_{p_j} \leq \left(\|p_j - r\|_\infty^{-\|p_j - r\|_\infty} + \|p_j - r\|_\infty |\Omega| \right) \|u - u_j\|_r,$$

valid for $p_j \leq r$ shows that

$$\|u - u_j\|_{p_j} \longrightarrow 0 \text{ as } j \longrightarrow \infty,$$

whereby in view of Proposition 1.5 one concludes that

$$\int_\Omega |u - u_j|^{p_j} \longrightarrow 0 \quad \text{as } j \longrightarrow \infty.$$

Invoking the inequalities (valid for $0 < t < 1$)

$$\int_\Omega |u_j|^{p_j} - \int_\Omega |u|^p \tag{4.49}$$

$$\leq \int_\Omega t^{1-p_+} |u_j - u|^{p_j} + (1-t)^{1-p_+} \int_\Omega |u|^{p_j} - \int_\Omega |u|^p$$

$$= \int_\Omega t^{1-p_+} |u_j - u|^{p_j} + (1-t)^{1-p_+} \left(\int_\Omega |u|^{p_j} - \int_\Omega |u|^p \right)$$

$$+ \left((1-t)^{1-p_+} - 1 \right) \int_\Omega |u|^p,$$

and

$$\int_\Omega |u|^p - \int_\Omega |u_j|^{p_j} \tag{4.50}$$

$$\leq \int_\Omega t^{1-p_+} |u - u_j|^p + (1-t)^{1-p_+} \int_\Omega |u_j|^p - \int_\Omega |u_j|^{p_j}$$

$$\leq \int_\Omega t^{1-p_+} |u_j - u|^p$$

$$+ (1-t)^{1-p_+} \left(\left(\|p - p_j\|_\infty^{-\|p - p_j\|_\infty} - 1 \right) \int_\Omega |u_j|^{p_j} + \|p - p_j\|_\infty |\Omega| \right)$$

$$+ \left((1-t)^{1-p_+} - 1 \right) \int_\Omega |u|^p,$$

recalling (4.36) and letting $j \to \infty$ one obtains the claim. This settles the case $p_+ < n$.

As for the case $n \in [p_-, p_+]$, we refer to the setting of the proof of Theorem 2.5: in particular there exists an open covering of Ω consisting of a finite collection of subdomains of Ω

$$\Omega_1, \Omega_2, ..., \ \Omega_M$$

and sequences of positive real numbers $s_1, ..., \ s_M$ and $\theta_1, ..., \ \theta_M$ with

$$p(x) + \theta_k \leq s_k \text{ in } \Omega_k.$$

Consider a partition of unity

$$\varphi_k, \quad 1 \leq k \leq M$$

subordinated to the open covering just described. By the assumption of uniform convergence, for sufficiently large I, there is a positive number θ such that for each $k = 1, 2, ... M$ one has:

$$p_I + \theta < s_k$$

uniformly in Ω. From the proof of Sobolev's embedding theorem (Theorem 2.5), we have the inclusion

$$(u_j\varphi_k) \subset L_{s_k}(\Omega)$$

and thus, for each $k = 1, 2, ... M$, the sequence

$$(u_j\varphi_k)$$

can be assumed to be convergent in $L_{s_k}(\Omega)$. Denote the limit of the sequence $(u_j\varphi_k)$ in $L_{s_k}(\Omega)$ by $v \in L_{s_k}(\Omega)$. In particular, for each k and for $j \geq I$, it holds that

$$u_j\varphi_k \in L_{p_I}(\Omega).$$

Moreover:

$$u_j \longrightarrow u \text{ in } L_p(\Omega),$$

and by reason of the inclusion

$$L_{s_k}(\Omega) \subseteq L_p(\Omega)$$

one has

$$\|u_j\varphi_k - v\|_p \to 0 \text{ as } j \longrightarrow \infty,$$

whence the inequality

$$\|v - \varphi_k u\|_p \leq \|u_j\varphi_k - v\|_p + \|(u_j - u)\varphi_k\|_p$$

implies that as $j \longrightarrow \infty$,

$$u_j\varphi_k \longrightarrow u\varphi_k \text{ in } L_{s_k}(\Omega) \hookrightarrow L_{p_I}(\Omega). \tag{4.51}$$

Consequently,

$$u = \sum_{k=1}^{M} u\varphi_k \in L_{p_I}(\Omega).$$

According to (4.51),

$$\|u - u_j\|_{p_I} \leq \sum_{k=1}^{M} \|(u - u_j)\varphi_k\|_{p_I} \to 0 \text{ as } j \to \infty;$$

from which it immediately follows that for $j \geq I$

$$\|u - u_j\|_{p_j} \longrightarrow 0 \text{ as } j \longrightarrow \infty,$$

whence

$$\int_\Omega |u - u_j|^{p_j} \longrightarrow 0 \text{ as } j \longrightarrow \infty.$$

Since

$$|u(x)|^{p_i(x)} \longrightarrow |u(x)|^{p(x)} \text{ as } i \longrightarrow \infty \text{ a.e. in } \Omega,$$

the pointwise estimate

$$|u|^{p_i} = \left| \sum_{k=1}^{M} u \varphi_k \right|^{p_i} \leq M^{p_i - 1} \sum_{k=1}^{M} |u \varphi_k|^{p_i} \leq C \sum_{k=1}^{M} |u \varphi_k|^{p_i}$$

and the pointwise estimate (valid in Ω for each $k = 1, 2 \ldots M$ and for any natural number $j \geq I$)

$$|u\varphi_k|^{p_j} \leq \chi_{\{x : |u\varphi_k(x)| > 1\}} |u|^{p_j} + \chi_{\{x : |u\varphi_k(x)| < 1\}} |u|^{p_j} \leq \chi_{\{|u(x)| > 1\}} |u|^{s_k} + 1,$$

imply easily via a straightforward application of Lebesgue's theorem that

$$\int_\Omega |u|^{p_j} \longrightarrow \int_\Omega |u|^{p} \text{ as } j \longrightarrow \infty.$$

On the other hand, it transpires from an application of inequalities (4.49) and (4.50) that

$$\int_\Omega |u_j|^{p_j} \longrightarrow \int_\Omega |u|^{p} \text{ as } j \to \infty.$$

Given the assumptions on p and the sequence (p_j) this implies that

$$\int_\Omega \frac{|u_j|^{p_j}}{p_j} \longrightarrow \int_\Omega \frac{|u|^{p}}{p} \text{ as } j \longrightarrow \infty. \tag{4.52}$$

Letting j tend to infinity in (4.48) we obtain

$$\limsup_{j} \int_\Omega \frac{|u_j|^{p_j}}{p_j} \leq \int_\Omega \frac{|u|^{p}}{p}.$$

In conjunction with inequality (4.48) this yields the maximality of u in the required sense.

If $p_- > n$ all the Sobolev spaces involved in the proof are contained in $L_\infty(\Omega)$. The sequence (u_j), in particular, fulfills the hypothesis of Lemma 4.8 for any real number $r > 1$. The maximality of u follows immediately from this observation and inequality (4.48). It is straightforward to show that (4.52) holds also in this case.

Next, owing to (4.52), for sufficiently large j,

$$\frac{p_-}{\int_\Omega |u|^p + 1} \leq \frac{\int_\Omega |\nabla u_j|^{p_j}}{\int_\Omega |u_j|^{p_j}} \leq \frac{p_{1,+}}{\int_\Omega |u|^p + 1};$$

there is therefore no loss of generality in assuming the sequence

$$\left(\frac{\int_\Omega |\nabla u_j|^{p_j}}{\int_\Omega |u_j|^{p_j}} \right)$$

to be convergent to, say, L. Clearly, from (4.52),

$$\int_\Omega |\nabla u_j|^{p_j} \longrightarrow L \int_\Omega |u|^p.$$

From weak lower semi-continuity it easily follows that

$$\int_\Omega |\nabla u|^p \leq \liminf_{j \to \infty} \int_\Omega |\nabla u_j|^p$$

$$\leq \liminf_{j \to \infty} \left(\|p_j - p\|_\infty^{-\|p-p_j\|} \int_\Omega |\nabla u_j|^{p_j} + \|p - p_j\|_\infty |\Omega| \right),$$

from which, utilizing (4.52), it is readily concluded that

$$\frac{\int_\Omega |\nabla u|^p}{\int_\Omega |u|^p} \leq L.$$

Inequalities (4.45) and (4.46) are proved along the same lines as the corresponding claims in Theorem 4.9.   □

In particular, if $p_- \geq 2$ the validity of Clarkson's inequality allows for a substantial improvement of the above discussion:

Theorem 4.13. *For $p_- \geq 2$, and u, $(u_i)_i$ as in Theorem 4.12, we have:*

$$\lim_{j \to \infty} \int_\Omega |\nabla u - \nabla u_j|^p \, dx = 0.$$

Proof. Notice that since $u \in \overset{\circ}{W}{}^1_p (\Omega)$, one has

$$\lim_{k \to \infty} \int_\Omega \frac{|\nabla u|^{p_k}}{p_k} dx = \int_\Omega \frac{|\nabla u|^p}{p} dx = 1.$$

On the other hand, the functional

$$G : \overset{\circ}{W}{}^1_p (\Omega) \to [0, \infty)$$

$$G(v) = \int_\Omega \frac{|\nabla v|^p}{p} dx$$

is weakly-lower semicontinuous, whence

$$1 = \int_\Omega \frac{|\nabla u|^p}{p} \le \liminf_{j \to \infty} \int_\Omega \frac{|\nabla u_j|^p}{p} \tag{4.53}$$

and

$$\int_\Omega \frac{|\nabla u|^p}{p} dx \le \liminf_j \int_\Omega \left| \frac{\nabla(u + u_j)}{2} \right|^p \frac{1}{p} dx.$$

From the equality (valid for all natural numbers j):

$$\int_\Omega \frac{|\nabla u_j|^{p_j}}{p_j} = 1$$

and the embedding estimate (4.22) one has, for $j \in \mathbb{N}$ and $\varepsilon = \|p_j - p\|_\infty$:

$$\int_\Omega \frac{|\nabla u_j|^p}{p} \le \varepsilon^{-\varepsilon} \int_\Omega \frac{|\nabla u_j|^{p_j}}{p^{p_j/p}} + \varepsilon|\Omega|$$

$$\le \varepsilon^{-\varepsilon} \int_\Omega |\nabla u_j|^{p_j} \left(\frac{1}{p^{p_j/p}} - \frac{1}{p_j} \right) + \varepsilon^{-\varepsilon} \int_\Omega \frac{|\nabla u_j|^{p_j}}{p_j} + \varepsilon|\Omega|$$

$$= \varepsilon^{-\varepsilon} \int_\Omega |\nabla u_j|^{p_j} \left(\frac{1}{p^{p_j/p}} - \frac{1}{p_j} \right) + \varepsilon^{-\varepsilon} + \varepsilon|\Omega|,$$

which implies

$$\limsup_{j \to \infty} \int_\Omega \frac{|\nabla u_j|^p}{p} \le 1. \tag{4.54}$$

Inequalities (4.53) and (4.54) yield

$$\lim_{j \to \infty} \int_\Omega \frac{|\nabla u_j|^p}{p} = 1. \tag{4.55}$$

Moreover,

$$\int_\Omega |\nabla u_j|^{p_j} \le p_{j+} \le p_+ \tag{4.56}$$

for all positive integers j and

$$\frac{1}{p^{p_j/p}} - \frac{1}{p_j} \longrightarrow 0 \text{ as } j \to \infty. \tag{4.57}$$

Consequently, for each $j \in \mathbb{N}$:

$$\int_\Omega \left| \frac{\nabla(u + u_j)}{2} \right|^p \frac{1}{p} dx \le \frac{1}{2} \int_\Omega \frac{|\nabla u|^p}{p} + \frac{1}{2} \int_\Omega \frac{|\nabla u_j|^p}{p}$$

$$\le \frac{1}{2} \int_\Omega \frac{|\nabla u|^p}{p} + \frac{1}{2} \left(\varepsilon^{-\varepsilon} \int_\Omega \frac{|\nabla u_j|^{p_j}}{p^{p_j/p}} + \varepsilon|\Omega| \right)$$

$$= \frac{1}{2} + \frac{1}{2} \left(\varepsilon^{-\varepsilon} \int_\Omega \frac{|\nabla u_j|^{p_j}}{p^{p_j/p}} + \varepsilon|\Omega| \right). \tag{4.58}$$

It follows that the expression to the right of the equal sign in (4.58) remains bounded as $j \to \infty$ and hence that

$$Q = \liminf_j \int_\Omega \left| \frac{\nabla(u + u_j)}{2} \right|^p \frac{1}{p} \, dx < \infty.$$

Fix $\delta > 0$ and select k so large that for $j \geq k$ one has

$$\|p - p_j\|_\infty |\Omega| < \delta$$

and

$$\|p - p_j\|_\infty^{-\|p - p_j\|_\infty} < 1 + \delta.$$

Pick I large enough so that $i \geq I$ implies

$$Q - \delta < \inf_{j \geq i} \int_\Omega \left| \frac{\nabla(u + u_j)}{2} \right|^p \frac{1}{p}.$$

Then for $j \geq I$;

$$\begin{aligned}
1 = \int_\Omega \frac{|\nabla u|^p}{p} dx &\leq \delta + \int_\Omega \left| \frac{\nabla(u + u_j)}{2} \right|^p \frac{1}{p} \\
&\leq \delta + \frac{1}{2}\left(1 + \int_\Omega \frac{|\nabla u_j|^p}{p} \right) \\
&\leq \delta + \frac{1 + \delta}{2}\left(1 + \int_\Omega \frac{|\nabla u_j|^{p_j}}{p^{p_j/p}} \right) \\
&\leq \delta + \frac{1 + \delta}{2}\left(1 + \int_\Omega |\nabla u_j|^{p_j} \left(\frac{1}{p^{p_j/p}} - \frac{1}{p_j} \right) + \int_\Omega \frac{|\nabla u_j|^{p_j}}{p_j} \right) \\
&= \delta + \frac{1 + \delta}{2}\left(1 + \int_\Omega |\nabla u_j|^{p_j} \left(\frac{1}{p^{p_j/p}} - \frac{1}{p_j} \right) + 1 \right).
\end{aligned}$$

By virtue of (4.56) and (4.57) it is apparent that the last term above tends to 1 as $j \to \infty$ and $\delta \longrightarrow 0$, whence

$$\int_\Omega \left| \frac{\nabla(u + u_j)}{2} \right|^p \frac{1}{p} \longrightarrow 1 \text{ as } j \longrightarrow \infty. \tag{4.59}$$

For any real number $t : t \geq 2$ the inequality

$$\left| \frac{a + b}{2} \right|^t + \left| \frac{a - b}{2} \right|^t \leq \frac{1}{2}\left(|a|^t + |b|^t \right)$$

holds for any complex numbers a and b (See [[4], Lemma 2.27]). In particular, for each $x \in \Omega$

$$\int_\Omega \left| \frac{\nabla(u + u_j)}{2p^{1/p}} \right|^p dx + \int_\Omega \left| \frac{\nabla(u - u_j)}{2p^{1/p}} \right|^p dx \leq$$

$$\frac{1}{2}\left(\int_\Omega \left| \frac{\nabla u}{p^{1/p}} \right|^p dx + \int_\Omega \left| \frac{\nabla u_j}{p^{1/p}} \right|^p dx \right)$$

whence (4.55) and (4.59) imply

$$\int_\Omega |\nabla(u - u_j)|^p dx \le 2^{p+} p^{p+/p-} \int_\Omega \left| \frac{\nabla(u - u_j)}{2p^{1/p}} \right|^p dx \longrightarrow 0 \text{ as } j \longrightarrow \infty.$$

This completes the proof of Theorem 4.13. □

The one-dimensional case deserves special consideration:

Theorem 4.14. *Consider* $\Omega = (a, b)$, $a < b$. *Let* $(p_j) \subset C([a, b])$ *be a non-increasing sequence of functions, uniformly convergent to* $p = \inf p_j$ *and assume*

$$1 < p_- \le p_+ < \infty.$$

For each $j \in \mathbb{N}$ *let* λ_j *and* u_j *be, respectively the first eigenvalue and a first eigenfunction of the problem*

$$\left(|u'|^{p_j - 2} u' \right) = \gamma |u|^{p_j - 2} u;$$

analogously, let λ *be the first modular eigenvalue of the* p-*Laplacian.*
Then there exists a subsequence (u_k) *of* (u_j) *and a (first) eigenfunction* u *of the* p-*Laplacian such that*

$$u_k \longrightarrow u \text{ in } \overset{\circ}{W}{}^1_p (\Omega).$$

Proof. For $p_- \ge 2$, Theorem 4.14 follows directly from Theorem 4.12. We hence focus on the case $p_- < 2$. To that end we notice the elementary fact that there exists a positive constant c such that as long as $1 < s < S < \infty$ and $(q, r) \in (s, S) \times [-M, M]$,

$$\frac{1}{2} \left(|r + 1|^q + |r - 1|^q \right) - |r|^q \ge c. \tag{4.60}$$

Set

$$A = \{x \in (a, b) : |(u_j - u)'| > \delta |(u_j + u)'|\}$$

and fix $\delta > 0$. Then if $x \in A$, $r = \frac{v(x) + v_j(x)}{v(x) - v_j(x)}$, $q = p_j(x)$, the substitution

$$v = \frac{u'}{p_j^{1/p_j}} \quad v_j = \frac{u_j'}{p_j^{1/p_j}}$$

in (4.60) yields, pointwise in Ω:

$$\frac{1}{2} \left(|v|^{p_j} + |v_j|^{p_j} \right) - \left| \frac{v + v_j}{2} \right|^{p_j} \ge c \left| \frac{v - v_j}{2} \right|^{p_j},$$

whence it follows by integrating on A that

$$\frac{1}{2}\left(\int_\Omega |v|^{p_j}\,dx + \int_\Omega |v_j|^{p_j}\,dx\right) - \int_\Omega \left|\frac{v+v_j}{2}\right|^{p_j}\,dx \geq c\int_A \left|\frac{v-v_j}{2}\right|^{p_j}\,dx.$$

$$(4.61)$$

As seen in Theorem 4.13

$$\int_\Omega \left|\frac{v+v_j}{2}\right|^{p_j}\,dx \longrightarrow 1 \text{ as } j \longrightarrow \infty,$$

whereas

$$\int_\Omega |v(x)|^{p_j}\,dx \longrightarrow 1 \text{ as } j \longrightarrow \infty.$$

Letting j tend to infinity in (4.61) it is clear that

$$\int_A \left|\frac{v-v_j}{2}\right|^{p_j}\,dx \longrightarrow 0 \text{ as } j \to \infty$$

which through an easy calculation yields :

$$\int_A |u' - u'_j|^p\,dx \longrightarrow 0 \text{ as } j \longrightarrow \infty.$$

On the other hand

$$\int_{\Omega\setminus A} \left|\frac{v-v_j}{2}\right|^{p_j}\,dx \leq \delta^{p-} \int_\Omega \left|\frac{v+v_j}{2}\right|^{p_j}\,dx \leq \delta^{p-}.$$

In all,

$$\int_\Omega |u' - u'_j|^p\,dx \longrightarrow 0 \text{ as } j \longrightarrow \infty.$$

□

4.5.1 *Alternative Rayleigh quotients*

Because of the lack of homogeneity inherent to the non-constant exponent case the consideration of the Rayleigh quotient

$$Q(u) = \frac{\||\nabla u|\|_p}{\|u\|_p}$$

leads, in contrast to the case for constant p, to a problem essentially different from that treated in the previous discussion. In this direction we highlight the work [47]. Therein, the minimization problem of Q is considered. In this direction, the authors obtain the following Theorem:

Theorem 4.15. *There exists a non-negative minimizer of the Rayleigh quotient*

$$Q(u) = \frac{\||\nabla u|\|_p}{\|u\|_p}$$

in $\overset{\circ}{W}{}^1_p(\Omega)$. Any such minimizer is a solution of the Euler-Lagrange equation

$$div\left(\left|\frac{\nabla u}{K}\right|^{p-2}\frac{\nabla u}{K}\right) + \frac{K}{k}S\left|\frac{u}{k}\right|^{p-2}\frac{u}{k} = 0,$$

where

$$K = \||\nabla u|\|_p \ , \ k = \|u\|_p \ and \ S = \frac{\int_\Omega \left|\frac{\nabla u}{K}\right|^p}{\int_\Omega \left|\frac{u}{k}\right|^p}.$$

Proof. The existence of a minimizer is obtained via a minimizing sequence which can be easily produced via the compactness of the Sobolev Embedding (Theorem 2.5). The Euler-Lagrange equations are derived from Theorem 4.3 and Theorem 1.14. We refer the interested reader to [47] for the details. □

The stability of the eigenvalues and eigenfunctions in this setting can also be treated via methods parallel to the ones explained in this Chapter.

Notes

The Dirichlet eigenvalues for the p-Laplacian with Dirichlet boundary condition were first studied in [43] via a Ljusternik-Schnirelman type approach. In the cited work a necessary condition is given for Poincare's inequality to hold in modular form (Theorem 3.4). In particular, in the one dimensional case the authors show that the validity of Poincarés inequality in modular form is equivalent to the monotonicity of the variable exponent p. A closely related problem, namely

$$-div\left(|\nabla u|^{p-2}\nabla u\right) = \lambda|u|^{q-2}u$$

has also been extensively studied: various existence results for appropriate p, q were given in [39], [42] and [65]. Yet another related eigenvalue problem involving non-standard growth condition is given by

$$-div\left(\left(|\nabla u|^{p_1-2} + |\nabla u|^{p_2-2}\right)\nabla u\right) = \lambda|u|^{q-2}u;$$

this problem was treated in [66]. Non-homogeneous type eigenvalue problems have also been studied; for example the inclusion of a so-called non-local term $\eta[u]$ leads to the problem

$$-\eta[u]div\left(\left(|\nabla u|^{p-2}\right)\nabla u\right) = \lambda f(x, u);$$

we refer the reader to [67] for the specifics. In the cited work, a continuous family of positive eigenvalues for the above problem was found, using suitable variational techniques.

An interesting question naturally arises, namely whether all eigenvalues of the p-Laplacian with Dirichlet boundary condition are of variational type. The negative answer was given by Binding and Rynne in [13], to which we refer for further details.

Neumann eigenvalues for variable p have been dealt with in [40] via Ljusternik-Schirelmann methods.

For constant exponent p it is well known (see [8]) that the first Dirichlet eigenvalue is simple. Whether or not this important result remains true for variable p is as of yet, unknown.

Chapter 5

Approximation on L_p Spaces

In this chapter we present a brief summary of some results about the behaviour of the approximation, Bernstein, Gelfand and Kolmogorov numbers for certain operators on variable exponent spaces. We start with the definition of some classical s-numbers, n-widths and their basic properties, to immediately embark on the study of the behaviour of these numbers for the particular cases of the Hardy operator and a Sobolev-type embedding.

5.1 s-numbers and n-widths

In this and in the next section, X and Y will denote Banach spaces, and in agreement with the terminology introduced in Chapter 1, B_X will denote the closed unit ball in X and I_X the identity map on X; $B(X,Y)$ will stand for the space of all bounded linear maps of X to Y, and we shall use $B(X)$ instead of $B(X,X)$. Given a closed linear subspace M of X, the embedding map of M into X will be denoted by J_M^X and the canonical map of X onto the quotient space $X \backslash M$ by Q_M^X.

5.1.1 s-numbers

Let $s : T \longmapsto (s_n(T))$ be a rule that assigns to every bounded linear operator acting between any pair of Banach spaces a sequence of non-negative numbers that has the following properties:

(S1) $\|T\| = s_1(T) \geq s_2(T) \geq \ldots 0.$

(S2) $s_n(S + T) \leq s_n(S) + \|T\|$ for $S, T \in B(X,Y)$ and all $n \in \mathbb{N}$.

(S3) $s_n(BTA) \leq \|B\| \, s_n(T) \, \|A\|$ whenever $A \in B(X_0, X)$, $T \in B(X,Y)$, $B \in B(Y, Y_0)$ and $n \in \mathbb{N}$.

(S4) $s_n(Id : l_2^n \to l_2^n) = 1$ for all $n \in \mathbb{N}$.

(S5) $s_n(T) = 0$ when rank $(T) < n$.

If Properties $S1 - S5$ are satisfied, we call $s_n(T)$ (or $s_n(T : X \to Y)$) the nth s-number of T. We say that $s_n(T)$ is the nth s-number of T in the "strict" sense when **(S4)** is replaced by

(S6) $s_n(Id : E \to E) = 1$ for every Banach space E with $\dim(E) \geq n$.

Evidently **(S6)** implies **(S4)**, and so for a given operator T the class of s-numbers is larger than that of "strict" s-numbers. More information about s-numbers, and those that are "strict", can be found in [73].

We now move on to presenting some important s-numbers and to exploring certain relations between them. We start by introducing the concept of modulus of injectivity and its counterpart, the modulus of surjectivity and by establishing some of their basic properties. The modulus of injectivity of an operator $T \in B(X, Y)$ is defined to be

$$j(T) := \sup\{\rho \geq 0 : \|Tx\|_Y \geq \rho \|x\|_X \text{ for all } x \in X\};$$

the modulus of surjectivity of T is, on the other hand

$$q(T) := \sup\{\rho \geq 0 : T(B_X) \supset \rho B_Y\}.$$

Lemma 5.1. *Let X, Y and Z be Banach spaces.*
(i) If $S, T \in B(X, Y)$, then:

$$q(S + T) \leq q(S) + \|T\| \text{ and } j(S + T) \leq j(S) + \|T\|.$$

(ii) If $T \in B(X, Z)$ and $S \in B(Z, Y)$, then:

$$q(ST) \leq q(S) \|T\| \text{ and } j(ST) \leq \|S\| j(T).$$

If additionally S is surjective, then

$$q(ST) \leq \|S\| q(T).$$

Finally, if T is surjective, then

$$j(ST) \leq j(S) \|T\|.$$

Proof. Since the inequalities involving j are evident, we will only prove inequality (i) for q. Assume that

$$q(S + T) > \|T\|$$

and

$$0 < \varepsilon < q(S + T) - \|T\|.$$

Set

$$\rho = q(S + T) - \varepsilon,$$

take $y \in B_Y$ and define a sequence $\{x_i\}$ of elements of X by induction as follows:
$$Sx_1 + Tx_1 = (\rho - \|T\|)y, \quad \|x_1\|_X \leq (\rho - \|T\|)/\rho,$$
$$Sx_{n+1} + Tx_{n+1} = Tx_n, \quad \|x_{n+1}\|_X \leq \|Tx_n\|_Y / \rho \qquad \text{for all } n \in \mathbb{N}.$$
Obviously
$$\|x_n\|_X \leq \left(\frac{\|T\|}{\rho}\right)^{n-1} \frac{\rho - \|T\|}{\rho} \qquad \text{for all } n \in \mathbb{N}.$$
Since $\|T\| < \rho$ it follows that the series $\sum_1^\infty x_n$ is convergent: let
$$x = \sum_1^\infty x_n.$$
Furthermore, by virtue of the condition $\|x\|_X \leq 1$ and since
$$Sx = (\rho - \|T\|)y,$$
one has the inclusion
$$S(B_X) \supset (\rho - \|T\|)B_Y.$$
This guarantees the following inequality, from which the proof of inequality (i) is immediate
$$q(S) \geq \rho - \|T\| = q(S + T) - \|T\| - \varepsilon. \qquad \square$$

Definition 5.1. Let $T \in B(X,Y)$ and $n \in \mathbb{N}$. Then the nth approximation (isomorphism) numbers of T are defined, respectively by:
$$a_n(T) = \inf\{\|T - F\| : F \in B(X,Y), \text{ rank } (T) < n\}$$
$$\text{and } i_n(T) = \sup\left\{\|A\|^{-1}\|B\|^{-1}\right\},$$
where the supremum is taken over all possible Banach spaces G with
$$\dim(G) \geq n$$
and maps $A \in B(Y,G)$, $B \in B(G,X)$ such that ATB is the identity on G. (We underline the fact that the validity of the definition of $i_n(T)$ follows from [73], Lemma 1.1.) The Gelfand, Bernstein, Kolmogorov, Mityagin, Weyl, Chang and Hilbert numbers of T are defined, respectively, by the following:
$$c_n(T) = \inf\left\{\|TJ_M^X\| : \text{codim } (M) < n\right\};$$
$$b_n(T) = \sup\left\{j\left(TJ_M^X\right) : \dim(M) \geq n\right\};$$
$$d_n(T) = \inf\left\{\|Q_N^Y T\| : \dim(N) < n\right\};$$
$$m_n(T) = \sup\left\{q\left(Q_N^Y T\right) : \text{codim } (N) \geq n\right\};$$
$$x_n(T) = \sup\left\{a_n(TA) : \|A : l_2 \to X\| \leq 1\right\};$$
$$y_n(T) = \sup\left\{a_n(BT) : \|B : Y \to l_2\| \leq 1\right\};$$
$$h_n(T) = \sup\left\{a_n(BTA) : \|A : l_2 \to X\|, \|B : Y \to l_2\| \leq 1\right\}.$$

The next lemma illustrates the reason for the introduction of the above numbers.

Lemma 5.2. *The approximation, isomorphism, Gelfand, Bernstein, Kolmogorov, Mityagin, Weyl, Chang and Hilbert numbers are s-numbers; the first five of them are strict s-numbers.*

Proof. We prove only **(S6)** for the approximation numbers. Suppose there is a Banach space E, with $\dim(E) \geq n$, such that

$$a_n(Id : E \to E) < 1.$$

Then there exists $A \in B(E)$ with rank $(A) < n$ and $\|Id - A\| < 1$. Since $A = Id - (Id - A)$ is invertible in $B(E)$ (via the Neumann series), we have rank $(A) \geq n$, which is clearly a contradiction. Thus **(S6)** holds.

All the claimed properties of the isomorphism numbers are clear, except for **(S2)**, which we now proceed to prove: Suppose that

$$i_n(S + T) > \|T\|$$

and let

$$\varepsilon \in (0, i_n(S + T) - \|T\|) .$$

Then there exists a Banach space G, with $\dim(G) \geq n$, and maps

$$A \in B(G, X), B \in B(Y, G)$$

such that

$$I_G = B(S + T)A$$

and

$$\|B\|^{-1} \|A\|^{-1} \geq i_n(S + T) - \varepsilon > \|T\| .$$

Hence $\|BTA\| < 1$, whence the operator

$$BSA = B(S + T)A - BTA = I_G - BTA$$

is invertible. Since $I_G = (I_G - BTA)^{-1} BSA$ and

$$\left\|(I_G - BTA)^{-1}\right\| \leq (1 - \|BTA\|)^{-1} ,$$

we obtain

$$i_n(S) \geq \left\|(I_G - BTA)^{-1} B\right\|^{-1} \|A\|^{-1} \geq (1 - \|BTA\|) \|B\|^{-1} \|A\|^{-1}$$
$$\geq \|B\|^{-1} \|A\|^{-1} - \|T\| \geq i_n(S + T) - \varepsilon - \|T\| .$$

Thus

$$i_n(S+T) \le i_n(S) + \|T\| + \varepsilon$$

and **(S2)** follows clearly. As it is obvious how to show that the Gelfand numbers are s-numbers, we skip the proof.

For the Bernstein numbers we prove only **(S3)** as the other conditions are quite obvious. Suppose that $\varepsilon \in (0, b_n(BTA))$. Then there is a subspace M_0 of X_0, with $\dim(M_0) \ge n$, for which

$$b_n(BTA) - \varepsilon \le j\left(BTAJ_{M_0}^{X_0}\right).$$

Denote by A_0 the restriction of A to M_0, viewed as a map to $M := A(M_0)$. Then

$$BTAJ_{M_0}^{X_0} = BTJ_M^X A_0 \text{ and } \|A_0\| \le \|A\|.$$

By Lemma 5.1 (ii) it follows that

$$0 < b_n(BTA) - \varepsilon \le j\left(BTJ_M^X A_0\right) \le \left\|BTJ_M^X\right\| j(A_0),$$

which implies that $j(A_0) > 0$. Hence A_0 is injective, so that $\dim(M) \ge n$. Since A_0 is surjective, Lemma 5.1 (ii) gives

$$b_n(BTA) - \varepsilon \le j\left(BTJ_M^X A_0\right) \le \|B\| j(TJ_M^X) \|A_0\| \le \|B\| b_n(T) \|A\|,$$

which establishes **(S3)**.

We omit proofs of the assertions regarding the remaining numbers as they are either trivial or similar to the proofs given above. $\square$

The next lemma highlights the significance of the approximation numbers.

Lemma 5.3. *The approximation numbers are the largest among the s-numbers.*

Proof. Suppose that $T, A \in B(X,Y)$, with rank $(A) < n$, for $n \in \mathbb{N}$, and let s_n be some s-number. Then

$$s_n(T) \le s_n(A) + \|T - A\| = \|T - A\|,$$

which gives us $a_n(T) \ge s_n(T)$. $\square$

Next we introduce the extension and lifting properties for Banach spaces; we illustrate the significance of these properties through Theorems 5.1 and 5.2.

Definition 5.2. A Banach space Y is said to have the extension property if, for every map S_0 that maps a subspace of an arbitrary Banach space X into Y, there exists $S \in B(X,Y)$ such that $\|S\| = \|S_0\|$ and $S_0 x = SJ_M^X x$ for every $x \in M$.

Definition 5.3. A Banach space X has the lifting property if, for every $S_0 \in B(X, Y/N)$, where N is any closed linear subspace of an arbitrary Banach space Y, and any $\varepsilon > 0$, there is a map $S \in B(X, Y)$ such that $\|S\| \leq (1 + \varepsilon) \|S_0\|$ and $S_0 x = Q_N^Y S x$ for all $x \in X$.

Theorem 5.1. *Let X be a Banach space having the lifting property. Then*

$$d_n(S) = a_n(S) \quad \text{for all } n \in \mathbb{N} \text{ and all } S \in B(X, Y).$$

Proof. Set $S \in B(X, Y)$. We simply have to show that $a_n(S) \leq d_n(S)$. Let $\varepsilon > 0$ and let N be a subspace of Y such that $\dim(N) < n$ and

$$\left\| Q_N^Y S \right\| \leq d_n(S) + \varepsilon.$$

Then there is a (lifting) map $T \in B(X, Y)$ for which

$$\|T\| \leq (1 + \varepsilon) \left\| Q_N^Y S \right\|$$

and

$$Q_N^Y S x = Q_N^Y T x \text{ for every } x \in X.$$

Define $A = S - T$. Then $A x \in N$ for all $x \in X$: rank $(A) < n$. Hence $a_n(S) \leq \|S - A\| = \|T\| \leq (1 + \varepsilon)(d_n(S) + \varepsilon)$, and then the result follows. $\square$

Theorem 5.2. *Let Y be a Banach space with the extension property. Then for each $n \in \mathbb{N}$ and each $S \in B(X, Y)$, one has*

$$c_n(S) = a_n(S).$$

Proof. Set $S \in B(X, Y)$. Since the approximation numbers are the largest s-numbers, it suffices to show that $a_n(S) \leq c_n(S)$. Let $\varepsilon > 0$. There is a subspace M of X, with codim $(M) < n$, such that

$$\left\| S J_M^X \right\| \leq c_n(S) + \varepsilon.$$

As Y has the extension property, there is an extension $T \in B(X, Y)$ of $S J_M^X$ with

$$\|T\| = \left\| S J_M^X \right\|.$$

Put $A = S - T$: then rank $(A) < n$ and $A x = 0$ for every $x \in M$. Hence

$$a_n(S) \leq \|S - A\| = \|T\| = \left\| S J_M^X \right\| \leq c_n(S) + \varepsilon,$$

which concludes the proof. $\square$

Lemma 5.4. *The isomorphism numbers are the smallest of the strict s-numbers and the Hilbert numbers are the smallest among all s-numbers.*

Proof. Let $T \in B(X, Y)$, $B \in B(G, X)$, $A \in B(Y, G)$ be such that $I_G = ATB$ and $\dim(G) \geq n$; let s_n be an arbitrary strict s-number. Then

$$1 = s_n(I_G) \leq \|A\| \, s_n(T) \, \|B\|,$$

from which we have $i_n(T) \leq s_n(T)$. The proof for the Hilbert numbers is a natural variation of this argument. $\qquad\square$

We next introduce the idea of injective and surjective s-numbers.

Definition 5.4. Let s_n be an s-number. Then
(i) s_n is called **surjective** if, for every quotient space $X \backslash M$ of X and every $S \in B(X \backslash M, Y)$, it holds that

$$s_n \left(S Q_M^X \right) = s_n(S);$$

(ii) s_n is said to be **injective** if, for every subspace N of Y and every $S \in B(X, N)$, one has

$$s_n \left(J_N^Y S \right) = s_n(S).$$

Lemma 5.5. *The largest surjective s-numbers are the Kolmogorov numbers; the smallest surjective numbers are the Mityagin numbers.*

Proof. It is plain that the Kolmogorov numbers are surjective. On the other hand, every Banach space X can be identified with a quotient space of some Banach space X^1 with the lifting property (see [74], C.3.7). Let Q_X^1 be the canonical map from X^1 onto X and consider $S \in B(X, Y)$. Then

$$d_n(S) = d_n(S Q_X^1) = a_n(S Q_X^1).$$

Let s_n be any surjective s-number. Then since the approximation numbers are the largest s-numbers, it follows that

$$s_n(S) = s_n(S Q_X^1) \leq a_n(S Q_X^1) = d_n(S).$$

The assertion about the Mityagin numbers follows from a modification of the proof given in the next Lemma concerning the Bernstein numbers. $\qquad\square$

Lemma 5.6. *The largest among injective s-numbers are the Gelfand numbers; the smallest injective strict s-numbers are the Bernstein numbers.*

Proof. It is clear that the Gelfand numbers are injective. We recall the fact that every Banach space Y is a subspace of a Banach space Y^∞ which has the extension property (see [74], C.3.3). Let $S \in B(X, Y)$;

then $c_n(S) = c_n\left(J_Y^\infty S\right)$, where J_Y^∞ is the embedding from Y to Y^∞. In conjunction with Theorem 5.2 this gives

$$c_n(S) = a_n\left(J_Y^\infty S\right). \tag{5.1}$$

Now let s_n be any injective s-number. Then, as can be easily seen to follow from Lemma 5.3,

$$s_n(S) = s_n\left(J_Y^\infty S\right) \le a_n\left(J_Y^\infty S\right).$$

In view of (5.1), this concludes the proof for the Gelfand numbers.

As to the Bernstein numbers, let $T \in B(X,Y)$, and consider an injective strict s-number s_n and a subspace M of X with $\dim\left(M\right) \ge n$; suppose $j\left(T J_M^X\right) > 0$. Put $M_0 = T(M)$ and denote by T_0 the restriction of T to M, viewed as a map from M to M_0. Then T_0 is invertible and

$$\left\|T_0^{-1}\right\| = j\left(T J_M^X\right)^{-1}.$$

Hence

$$1 = s_n(I_M) \le s_n(T_0)\left\|T_0^{-1}\right\| = s_n(J_{M_0}^Y T_0)\left\|T_0^{-1}\right\|$$
$$\le s_n(T J_M^X)\left\|T_0^{-1}\right\| \le s_n(T) j\left(T J_M^X\right)^{-1}.$$

Therefore, $j\left(T J_M^X\right) \le s_n(T)$, which readily yields

$$b_n(T) \le s_n(T). \qquad \square$$

In the next theorem, whose proof we omit, we present some connections between the s-numbers of a map $T \in B(X,Y)$ and those of its dual $T' \in B(Y^*, X^*)$. For the proofs of the assertions concerning the approximation numbers we refer the interested reader to [36] and [51]; the other claims can be found in [72] and [73].

Theorem 5.3. *Let $n \in \mathbb{N}$ and $T \in B(X,Y)$.*
  Then:

$$c_n(T) = d_n(T') \ and \ m_n(T) = b_n(T'),$$
$$y_n(T) = x_n(T') \ and \ x_n(T) = y_n(T'),$$
$$h_n(T) = h_n(T') \ and \ i_n(T) \le i_n(T'),$$
$$a_n(T') \le a_n(T) \le 5\, a_n(T').$$

Furthermore, if T is compact, then

$$a_n(T) = a_n(T') \ and \ d_n(T) = c_n(T').$$

Theorems 5.4, 5.5 and 5.6 follow directly from the results presented so far in this section:

Theorem 5.4. *Let $n \in \mathbb{N}$ and $T \in B(X, Y)$. Then:*

$$a_n(T) \geq d_n(T) \geq i_n(T)$$

and

$$a_n(T) \geq c_n(T) \geq b_n(T) \geq i_n(T) \geq h_n(T).$$

Theorem 5.5. *Let $T \in B(X, Y)$ and $n \in \mathbb{N}$. Then*
$$d_n(T) \geq b_n(T).$$

Proof. It follows straight from the definition of the Bernstein and Kolmogorov numbers, namely
$$b_n(T) = \sup \left\{ j \left(T J_M^X \right) : \dim(M) \geq n \right\},$$
$$d_n(T) = \inf \left\{ \left\| Q_N^Y T \right\| : \dim(N) < n \right\},$$
that it suffices to show that $\left\| Q_N^Y T \right\| \geq j \left(T J_M^X \right)$. Suppose that $j \left(T J_M^X \right) > 0$. Then $\dim(T(M)) \geq n$ and there exists $x \in M$ such that
$$\left\| Q_N^Y T x \right\|_Y = \| T x \|_Y = 1.$$
Since $\| T x \|_Y = j \left(T J_M^X \right) \| x \|_X$ and $\left\| Q_N^Y T x \right\|_Y \leq \left\| Q_N^Y T \right\| \| x \|_X$, the inequality follows. $\square$

Theorem 5.6. *Let $n \in \mathbb{N}$ and $T \in B(X, Y)$. Then*
$$c_n(T) \geq m_n(T).$$

Proof. The inequality follows from Theorems 5.3 and 5.5. Indeed:
$$c_n(T) = d_n(T') \geq b_n(T') = m_n(T). \qquad \square$$

Finally, Theorem 5.7 is obtained by combining all the preceding inequalities and Theorems in this section.

Theorem 5.7. *Suppose $T \in B(X, Y)$. Then $a_n(T), i_n(T), c_n(T), d_n(T)$ and $b_n(T)$ are strict s-numbers and we have, for each $n \in \mathbb{N}$:*

$$a_n(T) \geq \max \left(c_n(T), d_n(T) \right) \geq \min \left(c_n(T), d_n(T) \right)$$
$$\geq \min \left(b_n(T), m_n(T) \right) \geq i_n(T),$$

and
$$a_n(T) \geq \max \left(x_n(T), y_n(T) \right) \geq \min \left(x_n(T), y_n(T) \right) \geq h_n(T).$$

The approximation numbers are the largest s-numbers, the Hilbert numbers are the smallest s-numbers and the isomorphism numbers are the smallest strict s-numbers.

5.1.2   *n-widths*

In this section we introduce the concept of n-widths and study them in some depth. They play an important role in approximation theory in fact motivated the introduction of s-numbers.

Definition 5.5. Let A be a centrally symmetric subset of X (so that 0 belongs to A and if x belongs to A so does $-x$) and let $n \in \mathbb{N}$.
(i) The **linear n-width** of A with respect to X is defined to be

$$\delta_n(A, X) = \inf_{P_n} \sup_{x \in A} \|x - P_n(x)\|_X \,,$$

where the infimum is taken over all $P_n \in B(X, Y)$ with rank n, that is,

$$\dim P_n(X) = n.$$

If $P_n^\delta \in B(X, Y)$ has the properties that rank $P_n^\delta \leq n$ and

$$\delta_n(A, X) = \sup_{x \in A} \left\|x - P_n^\delta(x)\right\|_X \,,$$

it is called an **optimal** linear operator for $\delta_n(A, X)$.
(ii) The **Kolmogorov n-width** of A with respect to X is given by

$$\widetilde{d}_n(A, X) = \inf_{X_n} \sup_{x \in A} \inf_{y \in X_n} \|x - y\|_X \,,$$

where the infimum is taken over all n-dimensional subspaces X_n of X. Any subspace X_n^K of X, with dimension at most n, for which

$$\widetilde{d}_n(A, X) = \sup_{x \in A} \inf_{y \in X_n^K} \|x - y\|_X \,,$$

is said to be an **optimal** subspace for $\widetilde{d}_n(A, X)$.
(iii) The **Gelfand n-width** of A with respect to X is defined by

$$\widetilde{c}_n(A, X) = \inf_{L_n} \sup_{x \in A \cap L_n} \|x\|_X \,,$$

where the infimum is taken over all closed subspaces L_n of X of codimension at most n. Any subspace L_n^G of X, with codimension at most n, for which

$$\widetilde{c}_n(A, X) = \sup \left\{ \|x\|_X : x \in A \cap L_n^G \right\},$$

is called an **optimal** subspace for $\widetilde{c}_n(A, X)$.
(iv) The **Bernstein n-width** of A with respect to X is given by

$$\widetilde{b}_n(A, X) = \sup_{X_{n+1}} \sup \left\{ \lambda \geq 0 : X_{n+1} \cap (\lambda B_X) \subset A \right\},$$

where the outer supremum is taken over all subspaces X_{n+1} of X with dimension $n + 1$. Any subspace X_{n+1}^B of X, with dimension $n + 1$, for which $X_{n+1}^B \cap \left(\widetilde{b}_n(A, X) B_X \right) \subset A$, is said to be an **optimal** subspace for $\widetilde{b}_n(A, X)$.

In Lemmas 5.7 and 5.8, we describe some basic properties of the afore-mentioned widths.

Lemma 5.7. *Let A be a centrally symmetric subset of X, $n \in \mathbb{N}$ and assume that s_n stands for any of the numbers $\delta_n(A, X)$, $\widetilde{d}_n(A, X)$, $\widetilde{c}_n(A, X)$ or $\widetilde{b}_n(A, X)$. Then*

$$\sup_{a \in A} \|a\|_X \geq s_n(A, X) \geq s_{n+1}(A, X).$$

In addition, if Z is a Banach space such that $A \subset Z \subset X$, one has the inequality

$$s_n(A, X) \leq s_n(A, Z).$$

Proof. The proof is obvious. □

Lemma 5.8. *Let A be a centrally symmetric subset of X. Then A is relatively compact if and only if A is bounded and $\widetilde{d}_n(A, X) \downarrow 0$ as $n \uparrow \infty$.*

Proof. First suppose that $\overline{A}$ is compact. Accordingly, for every $\varepsilon > 0$ there is a finite ε-net of A, that is, a set of points $\{x_1, ..., x_n\}$ such that for every $x \in A$, one has

$$\min\{\|x - x_i\| : i = 1, ..., n\} \leq \varepsilon.$$

Considering $X_n := \mathrm{sp}\,\{x_1, ..., x_n\}$, it is clear from the definition of $\widetilde{d}_n(A, X)$ that $\widetilde{d}_n(A, X) \leq \varepsilon$; and since $\widetilde{d}_{n+1}(A, X) \leq \widetilde{d}_n(A, X)$, it follows from Lemma 5.7 that $\widetilde{d}_n(A, X) \downarrow 0$.

For the reverse implication, suppose that A is bounded and that $\widetilde{d}_n(A, X) \downarrow 0$ as $n \uparrow \infty$. Since A is bounded it is obvious that

$$\widetilde{d}_0(A, X) := \sup\{\|x\|_X : x \in A\} < \infty.$$

Let $\varepsilon > 0$. Then there exists $N \in \mathbb{N}$ such that $\widetilde{d}_n(A, X) < \varepsilon$ if $n \geq N$, and hence there is an n-dimensional subspace X_n of X with the property that

$$\sup_{a \in A} \inf_{y \in X_n} \|x - y\|_X < \varepsilon.$$

Thus to each $x \in A$ there corresponds a $y \in X_n$ with

$$\|x - y\|_X < \varepsilon,$$

so that $\|y\|_X < \widetilde{d}_0(A, X) + \varepsilon$. Since

$$\{y \in X_n : \|y\|_X \leq \widetilde{d}_0(A, X) + \varepsilon\}$$

is a compact subset of X_n, it has an ε-net, which is clearly a 2ε-net for A. The proof is complete. □

In particular, if $T \in B(X, Y)$ then setting $A = T(B_X)$ in Definition 5.5 one obtains the n-widths of T.

Definition 5.6. Let $T : X \to Y$ be linear (possibly unbounded) and let $n \in \mathbb{N}$. The nth linear width of T is defined as

$$\widetilde{\delta}_n(T) := \widetilde{\delta}_n(T(B_X), Y).$$

The nth Kolmogorov, Gelfand and Bernstein n-widths of T ($\widetilde{d}_n(T), \widetilde{c}_n(T)$ and $\widetilde{b}_n(T)$, respectively) are defined in the obvious analogous manner.

Remark 5.1. We highlight the difference between the Gelfand, Bernstein and Kolmogorov numbers of Definition 5.1 and the corresponding n-widths given by Definition 5.6. For example,

$$d_n(T) = \widetilde{d}_{n-1}(T).$$

We refer the interested reader to [75], Chapter 2 and [72], 6.2.6 for further details and comments on this point. These books are also excellent sources of information about the properties of s-numbers.

5.2   A Sobolev Embedding

We now focus on studying the s-numbers for Sobolev embeddings. Throughout this section $I = [a, b]$ will be a compact interval on the real line and we will suppose, generally, that $p \in \mathcal{P}(\Omega)$. By $W_p^1(I)$ and $\overset{\circ}{W}{}_p^1(I)$ we denote Sobolev spaces as they are introduced in Definition 2.1. Recall that $\overset{\circ}{W}{}_p^1(I)$ is the space of absolutely continuous functions on the interval I that vanish at the endpoints of I, and that $\overset{\circ}{W}{}_p^1(I)$ is equipped with the norm

$$f \longrightarrow \|f'\|_{p,I}$$

(note that $\|f'\|_{p,I}$ is a pseudonorm on $W_p^1(I)$).

We shall consider the following Sobolev embedding (with variable $p \in \mathcal{P}(\Omega)$):

$$E_0 : \overset{\circ}{W}{}_p^1(I) \to L_p(I), \tag{5.2}$$

and the norm of the embedding E_0 is defined by

$$\|E_0\| = \sup_{\|f'\|_{p,I} > 0, \ f(a) = f(b) = 0} \frac{\|f\|_{p,I}}{\|f'\|_{p,I}}.$$

We start by recalling a result for constant exponent $p \in (1, \infty)$. In that case the exact values of strict s-numbers for the Sobolev embedding (5.2) are known :

Theorem 5.8. *Let $n \in \mathbf{N}$ and $\widetilde{s}_n$ be any strict s-number. Then*

$$\widetilde{s}_n(E_0) = \gamma_p \frac{|I|}{n},$$

where $\gamma_p = (p')^{1/p} p^{1/p'} \pi^{-1} \sin(\pi/p)/2$.

Proof. See [31], Theorem 5.14. $\qquad\square$

Since I is bounded, the next Lemma follows from [72], Theorem V.4.18 or by adapting Theorems 2.3.1 and 2.3.4 in [27].

Lemma 5.9. *Let $r, s \in (1, \infty)$. Then E_0 is a compact map from $\overset{\circ}{W}{}_r^1(I)$ into $L_s(I)$.*

For non-constant p we have the next Lemma.

Lemma 5.10. *Let $1 < c < d < \infty$ and suppose that $p, q \in \mathcal{P}(I)$ are such that $p(x), q(x) \in (c, d)$ for all $x \in I$. Then $\overset{\circ}{W}{}_p^1(I)$ is compactly embedded into $L_q(I)$.*

Proof. By Theorem 1.12, $\overset{\circ}{W}{}_p^1(I)$ and $L_d(I)$ are continuously embedded in $\overset{\circ}{W}{}_c^1(I)$ and $L_q(I)$, respectively. By Lemma 5.9, $\overset{\circ}{W}{}_c^1(I)$ is compactly embedded into $L_d(I)$. The result now follows by composition of these maps. $\quad\square$

We next aim to determining the asymptotic behaviour of various s-numbers of E_0 when $p \in \mathcal{P}(I)$. To that effect, we introduce a variety of functions which will play a key role in estimating s-numbers for E_0.

Definition 5.7. Let $p, q \in \mathcal{P}(I)$, suppose that $J = (c, d) \subset I$ and let $\varepsilon > 0$; set

$$p_J^- = \inf\{p(x) : x \in J\}, \ p_J^+ = \sup\{p(x) : x \in J\}.$$

Then

$$A_{p,q}(J) := \inf_{y \in J} \sup \left\{ \|f\|_{q,J} : \|f'\|_{p,J} \leq 1, f(y) = 0 \right\},$$

$$B_p(J) := \inf_{y \in J} \sup \left\{ \|f\|_{p_J^+, J} : \|f'\|_{p_J^-, J} \leq 1, f(y) = 0 \right\},$$

$$C_{p,q}(J) := \sup\left\{\|f\|_{q,J} : \|f'\|_{p,J} \le 1, f(c) = f(d) = 0\right\}$$

and

$$D_p(J) := \sup\left\{\|f\|_{p_J^-,J} : \|f'\|_{p_J^+,J} \le 1, f(c) = f(d) = 0\right\}.$$

Definition 5.8. $N_{A_{p,q}}(\varepsilon)$ is the minimum of all those $n \in \mathbb{N}$ such that I can be written as

$$I = \bigcup_{j=1}^{n} I_j,$$

where each I_j is a closed sub-interval of I, $|I_i \cap I_j| = 0$ $(i \ne j)$ and

$$A_{p,q}(I_j) \le \varepsilon$$

for every j.

We define $N_{B_p}(\varepsilon)$, $N_{C_{p,q}}(\varepsilon)$ and $N_{D_p}(\varepsilon)$ in a similar way.

For the sake of brevity we shall write

$$A_p(J) = A_{p,p}(J)$$

and

$$C_p(J) = C_{p,p}(J),$$

or simply

$$A_p(J), C_p(J).$$

We will occasionally write

$$A_{p,q}(J) = A_{p,q}(J)$$

and

$$C_{p,q}(J) = C_{p,q}(J).$$

By a slight modification of [31], Theorem 5.8 and [31], Lemma 9.6 we have the next result for the case of constant p and q.

Lemma 5.11. *Let* $p, q \in (1, \infty)$ *and* $J = (c, d) \subset I$. *Then*

$$\mathfrak{B}(p,q)\,|J|^{1/p'+1/q} := \frac{(p'+q)^{1/p-1/q}(p')^{1/q}q^{1/p'}}{2B(1/p',1/q)}\,|J|^{1/p'+1/q}$$
$$= A_{p,q}(J) = C_{p,q}(J), \tag{5.3}$$

and

$$A_{p,p}(J) = B_{p,p}(J) = C_{p,p}(J) = \gamma_p|J|,$$

where $\gamma_p = p'^{1/p}p^{1/p'}\pi^{-1}\sin(\pi/p)/2$.

Lemma 5.12 establishes some basic properties of the above quantities:

Lemma 5.12. *Let $p, q \in \mathcal{P}(I)$ and suppose that $(c, d) \subset I$. Then the functions $A_{p,q}(c, t), B_p(c, t), C_{p,q}(c, t)$ and $D_p(c, t)$ of the variable t are nondecreasing and continuous; so are $A_{p,q}(t, d), B_p(t, d), C_{p,q}(t, d)$ and $D_p(t, d)$.*

Proof. First we prove that for $h \geq 0$ one has

$$A_{p,q}(c, d) \leq A_{p,q}(c, d + h).$$

Clearly

$$A_{p,q}(c, d + h) = \inf_{y \in (c, d+h)} \sup \left\{ \|f\|_{q,(c,d+h)} : \|f'\|_{p,(c,d+h)} \leq 1, f(y) = 0 \right\}$$
$$= \min\{X, Y\},$$

where

$$X = \inf_{y \in (c,d)} \sup \left\{ \|f\|_{q,(c,d+h)} : \|f'\|_{p,(c,d+h)} \leq 1, f(y) = 0 \right\}$$

and

$$Y = \inf_{y \in (d,d+h)} \sup \left\{ \|f\|_{q,(c,d+h)} : \|f'\|_{p,(c,d+h)} \leq 1, f(y) = 0 \right\}.$$

Now

$$X \geq \inf_{y \in (c,d)} \sup \left\{ \|f\|_{q,(c,d)} : \|f'\|_{p,(c,d)} \leq 1, f(y) = 0 \right\} = A(c, d)$$

and

$$Y \geq \inf_{y \in (d,d+h)} \sup \left\{ \|f\|_{q,(c,d)} : \|f'\|_{p,(c,d)} \leq 1, f(y) = 0 \right\}$$
$$\geq \sup \left\{ \|f\|_{q,(c,d)} : \|f'\|_{p,(c,d)} \leq 1, f(d) = 0 \right\}$$
$$\geq \inf_{y \in (c,d)} \sup \left\{ \|f\|_{q,(c,d)} : \|f'\|_{p,(c,d)} \leq 1, f(y) = 0 \right\} = A_{p,q}(c, d),$$

which gives $A_{p,q}(c, d + h) \geq A_{p,q}(c, d)$.

Next, we prove the continuity of A. By Hölder's inequality (Theorem 1.5) we have, for some $\alpha \geq 1$ (independent of f, x and y),

$$|f(x) - f(y)| \leq \alpha \|1\|_{p',(y,x)} \|f'\|_{p,(y,x)},$$

and considering $\|1\|_{p',(y,x)}$ as a function of x we obtain

$$\left\| \|1\|_{p'(\cdot),(y,x)} \right\|_{q,(d,d+h)} \leq \|1\|_{p',(c,d+h)} \|1\|_{q,(d,d+h)},$$

which gives

$$A_{p,q}(c, d) \leq A_{p,q}(c, d + h)$$
$$= \inf_{y \in (c,d+h)} \sup \left\{ \|f\|_{q,(c,d+h)} : \|f'\|_{p,(c,d+h)} \leq 1, f(y) = 0 \right\}.$$

With the understanding that, unless otherwise specified, the suprema are taken over all f with $\|f'\|_{p,(c,d+h)} \leq 1$ and $f(y) = 0$, we have

$$A_{p,q}(c, d + h) \leq \inf_{y \in (c,d+h)} \sup \left\{ \|f\|_{q,(c,d)} + \|f\|_{q,(d,d+h)} \right\}$$

$$\leq \inf_{y \in (c,d+h)} \sup \left\{ \|f\|_{q,(c,d)} + \alpha \left\| \|1\|_{p',(y,x)} \|f'\|_{p,(y,x)} \right\|_{q,(d,d+h)} \right\}$$

$$\leq \inf_{y \in (c,d+h)} \sup \left\{ \|f\|_{q,(c,d)} + \alpha \left\| \|1\|_{p',(y,x)} \right\|_{q,(d,d+h)} \right\}$$

$$\leq \inf_{y \in (c,d+h)} \sup \|f\|_{q,(c,d)} + \alpha \|1\|_{p',(c,d+h)} \|1\|_{q,(d,d+h)}$$

$$\leq \inf_{y \in (c,d)} \sup \|f\|_{q,(c,d)} + \alpha \|1\|_{p',(c,d+h)} \|1\|_{q,(d,d+h)}$$

$$\leq \inf_{y \in (c,d)} \sup \left\{ \|f\|_{q,(c,d)} : \|f'\|_{p,(c,d)} \leq 1 \right\}$$
$$+ \alpha \|1\|_{p',(c,d+h)} \|1\|_{q,(d,d+h)}$$

$$= A_{p,q}(c, d) + \alpha \|1\|_{p',(c,d+h)} \|1\|_{q,(d,d+h)}.$$

From $q \in \mathcal{P}(I)$ have $\|1\|_{q,(d,d+h)} \to 0$ as $h \to 0$, and so $A_{p,q}(c, \cdot)$ is right-continuous. Left-continuity is proved in an analogous manner and the continuity of $A_{p,q}(c, \cdot)$ follows thereafter. Essentially identical arguments can be used to prove the remaining statements. □

From Lemma 5.12 and compactness (guaranteed by Lemma 5.10) one obtains Lemma 5.13:

Lemma 5.13. *Let $p, q \in \mathcal{P}(I)$ and assume that $p_-, p_+, q_-, q_+ \in (1, \infty)$. Then*

$$E_0 : \overset{\circ}{W}{}^1_p(I) \to L_p(I)$$

is compact and for all $\varepsilon > 0$ the quantities $N_{A_p}(\varepsilon)$, $N_{B_p}(\varepsilon)$, $N_{C_{p,q}}(\varepsilon)$ and $N_{D_p}(\varepsilon)$ are finite.

Moreover:

Lemma 5.14. *Let $p \in \mathcal{P}(I)$ and write $A = A_p$. Then for any $N \in \mathbb{N}$, there exists a unique $\varepsilon > 0$ such that $N_A(\varepsilon) = N$, and there is a covering of I by non-overlapping intervals I^i_A ($i = 1, ..., N$) such that $A(I^i_A) = \varepsilon$ for $i = 1, ..., N$. The same holds when A is replaced by B, C, D.*

Proof. The existence of the stipulated $\varepsilon > 0$ follows easily from the continuity properties established in Lemma 5.12. On the other hand, given two coverings of I, say $\{I_A^i\}_{i=1}^N$ and $\{J_A^i\}_{i=1}^N$, there are m, j, k, l such that $I_A^m \subset J_A^j$ and $J_A^k \subset I_A^l$. Assuming that $A(I_A^i) = \varepsilon_1$ and $A(J_A^i) = \varepsilon_2$, we obtain $\varepsilon_1 \leq \varepsilon_2 \leq \varepsilon_1$ by the monotonicity of A and this guarantees the uniqueness claim. $\qquad\square$

5.2.1 *The case when p is a step-function*

Next, we consider the case when the exponent is a step function. Consider p to be of the form:

$$p(x) = \sum_{i=1}^m \chi_{J_i}(x) p_i, \qquad (5.4)$$

where $\{J_i\}_{i=1}^m$ is a covering of I by non-overlapping intervals and $p_i \in (1, \infty)$. In the interest of clarity, we might write A instead of A_p; similarly B, C, D will be used for the corresponding quantities introduced before.

Lemma 5.15. *Let p be the step-function given by (5.4). Then, the map*

$$E_0 : \overset{\circ}{W}{}_p^1(I) \to L_p(I)$$

is compact. For $\varepsilon > 0$ small enough we have:
  (i) $b_{N_C(\varepsilon)-m}(E_0) > \varepsilon$ *and*
  (ii) $a_{N_A(\varepsilon)+2m-1}(E_0) < \varepsilon$.

Proof. The compactness of E_0 follows from Lemma 5.13; for $\varepsilon > 0$ it is easy to see that $N_A(\varepsilon)$ and $N_C(\varepsilon)$ are finite.
  (i) By virtue of the continuity of $C(c, \cdot)$, there is a set

$$\{I_i : i = 1, ..., N_C(\varepsilon)\}$$

of non-overlapping intervals covering I and such that $C(I_i) = \varepsilon$ whenever

$$1 \leq i < N_C(\varepsilon)$$

and such that

$$C\left(I_{N_C(\varepsilon)}\right) \leq \varepsilon.$$

Let $\eta \in (0, \varepsilon)$. Then, corresponding to each i with $1 \leq i < N_C(\varepsilon)$, there is a function f_i such that

$$\operatorname{supp} f_i \subset I_i := (a_i, a_{i+1}),$$

with the property that

$$\|f_i'\|_p = 1 \ , \ \varepsilon - \eta < \|f_i\|_p \le \varepsilon$$

and with

$$f(a_i) = f(a_{i+1}) = 0.$$

Denote by

$$\{I_{i_k}\}_{k=1}^M$$

the set of those intervals I_i, $1 \le i < N_C(\varepsilon)$, each of which is contained in one of the intervals J_l given in the definition of p in (5.4). Then we have:

$$N_C(\varepsilon) - m \le M \le N_C(\varepsilon).$$

Observe that the set

$$X_M = \left\{ f' = \sum_{r=1}^M \alpha_{i_r} f_{i_r}' : \alpha_{i_r} \in \mathbb{R} \right\}$$

is an M-dimensional subspace of $W_p^1(I)$. Note that since p is constant on I_{i_r}, $p(x) = p_{i_r}$ on I_{i_r}. Choose $f \in X_M \backslash \{0\}$. Setting $\lambda_0 := \|f\|_{p(\cdot)}$ it follows that

$$1 \ge \int_I \left| \frac{f(x)}{\lambda_0} \right|^{p(x)} dx \ge \sum_{r=1}^M \int_{I_{i_r}} \left| \frac{f(x)}{\lambda_0} \right|^{p(x)} dx$$

$$= \sum_{r=1}^M \left(\frac{1}{\lambda_0} \right)^{p_{i_r}} \int_{I_{i_r}} |f(x)|^{p_{i_r}} dx \ge \sum_{r=1}^M \left(\frac{\varepsilon - \eta}{\lambda_0} \right)^{p_{i_r}} \int_{I_{i_r}} |f'(x)|^{p_{i_r}} dx$$

$$= \sum_{r=1}^M \int_{I_{i_r}} \left| \frac{f'(x)}{\lambda_0/(\varepsilon - \eta)} \right|^{p(x)} dx = \int_{\cup_{r=1}^M I_{i_r}} \left| \frac{f'(x)}{\lambda_0/(\varepsilon - \eta)} \right|^{p(x)} dx$$

$$= \int_I \left| \frac{f'(x)}{\lambda_0/(\varepsilon - \eta)} \right|^{p(x)} dx.$$

Hence

$$\|f'\|_{p,I} \le \|f\|_{p,I} / (\varepsilon - \eta),$$

and so

$$b_{N_C(\varepsilon)-m}(E_0) \ge b_M(E_0) \ge \varepsilon - \eta.$$

(ii) The proof of this statement goes along lines similar to those of (i). In this case we consider a set of non-overlapping intervals $\{I_i\}_{i=1}^{N_A(\varepsilon)}$, covering I for which

$$A(I_i) = \varepsilon \quad \text{when} \quad i = 1, ..., N_A(\varepsilon) - 1$$

and such that $A\left(I_{N_A(\varepsilon)}\right) \le \varepsilon$. By $\{I_i^+\}_{i=1}^M$ we denote the family of all non-empty intervals for which there exist j and k such that $I_i^+ = I_j \cap I_k$. Clearly

$$N_A(\varepsilon) \le M \le N_A(\varepsilon) + 2(m-1).$$

Let $1 > \eta > 0$. Then given any $i \in \{1, ..., M\}$, there exists $y_i \in I_i^+$ such that

$$\sup\left\{\|f\|_{p,I_i^+} : \|f\|_{p,I_i^+} = 1, f(y_i) = 0\right\} \le \varepsilon + \eta.$$

Define

$$P_\varepsilon(f) = \sum_{i=1}^M f(y_i)\chi_{I_i^+}.$$

Obviously, P_ε is a linear map from $\overset{\circ}{W}{}_p^1(I)$ to $L_p(I)$ with rank M. Let p_i be the constant value of p on I_i^+. Then for any $\lambda_0 \in (0,\infty)$ and $f \in \overset{\circ}{W}{}_p^1(I)$,

$$\int_I \left|\frac{(E_0 - P_\varepsilon)f(x)}{\lambda_0}\right|^{p(x)} dx = \sum_{i=1}^M \int_{I_i^+} \left|\frac{f(x) - f(y_i)}{\lambda_0}\right|^{p(x)} dx$$

$$= \sum_{i=1}^M \lambda_0^{-p_i} \int_{I_i^+} |f(x) - f(y_i)|^{p_i} dx$$

$$\le \sum_{i=1}^M \lambda_0^{-p_i}(\varepsilon + \eta)^{p_i} \int_{I_i^+} |f'(x)|^{p_i} dx$$

$$= \int_I \left|\frac{f'(x)}{\lambda_0/(\varepsilon + \eta)}\right|^{p(x)} dx.$$

Now choose $\lambda_0 = (1-\eta)\|(E_0 - P_\varepsilon)f\|_{p,I}$. Then

$$1 < \int_I \left|\frac{(E_0 - P_\varepsilon)f(x)}{\lambda_0}\right|^{p(x)} dx \le \int_I \left|\frac{f'(x)}{\lambda_0/(\varepsilon + \eta)}\right|^{p(x)} dx,$$

from which we see that

$$\|f'\|_{p,I} > (1-\eta)\|(E_0 - P_\varepsilon)f\|_{p,I}/(\varepsilon + \eta),$$

so that

$$\frac{\varepsilon + \eta}{1 - \eta} > \frac{\|(E_0 - P_\varepsilon)f\|_{p,I}}{\|f'\|_{p,I}}.$$

Now let $\eta \to 0$ and the lemma follows. $\qquad\square$

Lemma 5.16. *Let p be the step-function given by (5.4). Then*

$$\lim_{\varepsilon \to 0} \varepsilon N(\varepsilon) = \frac{1}{2\pi} \int_I \left(p'(x) p(x)^{p(x)-1} \right)^{1/p(x)} \sin(\pi/p(x)) dx,$$

where N stands for N_A, N_B, N_C or N_D.

Proof. Use the fact that p is a step function together with Lemmas 5.11 and 5.14. $\qquad\square$

Finally, we arrive at the main result for the case when p is a step-function.

Theorem 5.9. *Let p be the step-function given by (5.4). Then for the compact map $E_0 : \overset{\circ}{W}{}^1_p(I) \to L_p(I)$ we have*

$$\lim_{n \to \infty} n s_n(E_0) = \frac{1}{2\pi} \int_I \left(p'(x) p(x)^{p(x)-1} \right)^{1/p(x)} \sin(\pi/p(x)) dx,$$

where $s_n(E_0)$ denotes the nth approximation, Gelfand, Kolmogorov or Bernstein number of E_0.

Proof. From Lemma 5.15 together with Theorems 5.4 and 5.5 it follows that

$$\varepsilon N_A(\varepsilon) \geq a_{N_A(\varepsilon)+2m-1}(E_0) N_A(\varepsilon) \geq b_{N_A(\varepsilon)+2m-1}(E_0) N_A(\varepsilon)$$

and that

$$\varepsilon N_C(\varepsilon) \leq b_{N_C(\varepsilon)-m}(E_0) N_C(\varepsilon).$$

A straightforward application of Lemma 5.16 now yields the result for the approximation and Bernstein numbers. The remaining assertions follow from the inequalities given in Theorems 5.4 and 5.5. $\qquad\square$

5.2.2 The case of strongly log-Hölder-continuous exponent

We aim to obtain a result similar to that of Theorem 5.9 by approximating p by step-functions. Corollary 1.6 allows for estimates of the changes in the various norms occurring when p is replaced by an approximating function. We start with some technical lemmas.

Lemma 5.17. *Let $\delta > 0$, let $J \subset I$ be an interval and suppose that $p, q \in \mathcal{P}(J)$ are such that*

$$p(x) \leq q(x) \leq p(x) + \delta \text{ for all } x \in J.$$

Then

$$\left(\delta |J| + \delta^{-\delta} \right)^{-2} A_{p+\delta,p}(J) \leq A_q(J) \leq \left(\delta |J| + \delta^{-\delta} \right)^2 A_{p,p+\delta}(J).$$

Proof. Set

$$B_1 = \{f : \|f\|_q \le 1\}, B_2 = \{f : \|f\|_p \le \delta |J| + \delta^{-\delta}\},$$

where the norms are with respect to the interval J. By Theorem 1.13, it is clear that

$$\|f\|_p \le \left(\delta |J| + \delta^{-\delta}\right) \|f\|_q,$$

which gives $B_1 \subset B_2$ and that

$$\begin{aligned}
A_q(J) &= \inf_{y \in J} \sup \left\{ \|f(\cdot) - f(y)\|_q : \|f'\|_q \le 1 \right\} \\
&= \inf_{y \in J} \sup \left\{ \|f(\cdot) - f(y)\|_q : f' \in B_1 \right\} \\
&\le \inf_{y \in J} \sup \left\{ \left(\delta |J| + \delta^{-\delta}\right) \|f(\cdot) - f(y)\|_{p+\delta} : f' \in B_2 \right\} \\
&= \left(\delta |J| + \delta^{-\delta}\right)^2 \inf_{y \in J} \sup \left\{ \left\| \frac{f(\cdot) - f(y)}{\delta |J| + \delta^{-\delta}} \right\|_{p+\delta} : \left\| \frac{f'}{\delta |J| + \delta^{-\delta}} \right\|_p \le 1 \right\} \\
&= \left(\delta |J| + \delta^{-\delta}\right)^2 \inf_{y \in J} \sup \left\{ \|g(\cdot) - g(y)\|_{p+\delta} : \|g'\|_p \le 1 \right\} \\
&= \left(\delta |J| + \delta^{-2}\right)^{-2} A_{p,p+\delta}(J).
\end{aligned}$$

The proof of the remaining part of the claimed inequality is similar. $\quad\square$

Lemma 5.18. *Let $J \subset I$ be an interval with $|J| \le 1$ and suppose that $\widetilde{p} \in (1, \infty)$. Then there is a bounded positive function η defined on $(0, 1)$, with $\eta(\delta) \to 0$ as $\delta \to 0$, such that if $p, q \in \mathcal{P}(J)$ with*

$$\widetilde{p} \le p(x) \le \widetilde{p} + \delta, \ \widetilde{p} \le q(x) \le \widetilde{p} + \delta \text{ in } J,$$

then

$$(1 - \eta(\delta)) |J|^{2\delta} \le \frac{A_p(J)}{A_q(J)} \le (1 + \eta(\delta)) |J|^{-2\delta}.$$

Proof. We prove only the right-hand inequality as the rest follows in a similar fashion. By Lemma 5.11 and (5.3) we have

$$\begin{aligned}
\frac{A_p(J)}{A_q(J)} &\le \left(\delta |J| + \delta^{-\delta}\right)^4 \frac{A_{\widetilde{p},\widetilde{p}+\delta}(J)}{A_{\widetilde{p}+\delta,\widetilde{p}}(J)} \\
&= \left(\delta |J| + \delta^{-\delta}\right)^4 \frac{\mathfrak{B}(\widetilde{p}, \widetilde{p} + \delta)}{\mathfrak{B}(\widetilde{p} + \delta, \widetilde{p})} |J|^{-2\delta/(\widetilde{p}(\widetilde{p}+\delta))} \\
&\le \left(\delta |J| + \delta^{-\delta}\right)^4 \frac{\mathfrak{B}(\widetilde{p}, \widetilde{p} + \delta)}{\mathfrak{B}(\widetilde{p} + \delta, \widetilde{p})} |J|^{-2\delta}.
\end{aligned}$$

Since

$$\lim_{\delta \to 0} \left(\delta \, |J| + \delta^{-\delta} \right)^4 \frac{\mathfrak{B}(\widetilde{p}, \widetilde{p} + \delta)}{\mathfrak{B}(\widetilde{p} + \delta, \widetilde{p})} = 1,$$

the choice

$$\eta(\delta) = \max \left\{ \delta, \left(\delta \, |J| + \delta^{-\delta} \right)^4 \frac{\mathfrak{B}(\widetilde{p}, \widetilde{p} + \delta)}{\mathfrak{B}(\widetilde{p} + \delta, \widetilde{p})} - 1 \right\}$$

establishes the lemma. ☐

In what follows we shall need a restriction on the function $p \in \mathcal{P}(I)$ that is a little stronger than the log-Hölder condition (see Definition 2.2). We remind the reader that $[a, b]$ is a compact interval.

Definition 5.9. A function $p \in \mathcal{P}(I)$ is said to be strongly log-Hölder continuous (written $p \in \mathcal{SLH}(I)$) if there is an increasing continuous function ψ defined on $[0, |I|]$ such that $\lim_{t \to 0+} \psi(t) = 0$ and

$$- |p(x) - p(y)| \log |x - y| \leq \psi \left(|x - y| \right)$$

$$\text{for all } x, y \in I \text{ with } 0 < |x - y| < 1/2. \tag{5.5}$$

It is easy to see that Lipschitz or Hölder functions belong to $\mathcal{SLH}(I)$.

Proposition 5.1. *Let* $p \in \mathcal{SLH}(I)$. *Then*

$$\lim_{\varepsilon \to 0} \varepsilon N(\varepsilon) = \frac{1}{2\pi} \int_I \left(p'(x) p(x)^{p(x)-1} \right)^{1/p(x)} \sin \left(\pi/p(x) \right) dx,$$

where N *stands for* N_{A_p} *or* N_{C_p}.

Proof. We prove only the case $N = N_{A_p}$, since the other case is similar. Let $N \in \mathbb{N}$. By Lemma 5.14, there are a constant $\varepsilon_N > 0$ and a set of non-overlapping intervals $\{I_i^N\}_{i=1}^N$ covering I such that $A_p(I_i^N) = \varepsilon_N$ for every i. Let q_N be the step-function defined by

$$q_N(x) = \sum_{i=1}^N p_{I_i^N}^+ \chi_{I_i^N}(x)$$

and set

$$\delta_{N,i} = p_{I_i^N}^+ - p_{I_i^N}^-.$$

Then

$$p(x) \leq q_N(x) \leq p(x) + \delta_{N,i} \text{ for } i = 1, ..., N.$$

Claim 1. $\varepsilon_N \to 0$ as $N \to \infty$.

To prove this, note that ε_N is non-increasing. Suppose that there exists $\delta > 0$ such that $\varepsilon_N > \delta$ for all N. Fix N and set $I_i^N = I_i = (a_i, a_{i+1})$. Since $A_{p,I_i} > \delta$, for each $i \in \{1, ..., N\}$ there is a function f_i, with $f_i' = 0$ outside of I_i, $f(y_i) = 0$ for some $y_i \in I_i$, $\|f_i'\|_{p,I_i} \leq 1$, and $\|f_i\|_{p,I_i} = \inf_c \|f_i - c\|_{p,I_i} > \delta$.

We have

$$\|f_i - f_j\|_{p(.),I} \geq \|f_i - f_j\|_{p(.),I_i} \geq \inf_c \|f_i - c\|_{p(.),I_i} > \delta \text{ for } i < j,$$

and so there are N functions $f_1, ..., f_N$ in the unit ball of $L_{p,I}$ such that

$$\|E_0(f_i - f_j)\|_{p,I_i} > \delta \text{ for } i \neq j.$$

Since N can be arbitrarily large, this contradicts the compactness of T and establishes the claim.

Claim 2. $\lim_{N \to \infty} \max \{ |I_i^N| : i = 1, 2, ..., N \} = 0.$

If this were false, there would be sequences $\{N_k, i_k\}_{k=1}^{\infty}$,

$$i_k \in \{1, 2, ..., N_k\}$$

and an interval J such that $J \subset I_{i_k}^{N_k}$ for each k, so that

$$\varepsilon_{N_k} = A_p \left(I_{i_k}^{N_k} \right) \geq A_p (J) > 0,$$

contradicting the fact that $\varepsilon_N \to 0$.

Claim 3. We start by proving that there is a sequence $\{\beta_N\}$, with $\beta_N \downarrow 1$, such that for all $i \in \{1, ..., N\}$,

$$\beta_N^{-1} \varepsilon_N \left| I_i^N \right|^{2\delta_{N,i}} \leq A_{q_N(\cdot)} \left(I_i^N \right) \leq \beta_N \varepsilon_N \left| I_i^N \right|^{-2\delta_{N,i}}.$$

This can be established by noting that since $p^- \leq q_N(x), p(x) \leq p^- + \delta_{N,i}$ on I_i^N, Lemma 5.18 implies that

$$(1 - \eta(\delta_{N,i})) \left| I_i^N \right|^{2\delta_{N,i}} \leq \frac{A_p \left(I_i^N \right)}{A_{q_N(\cdot)} \left(I_i^N \right)} \leq (1 + \eta(\delta_{N,i})) \left| I_i^N \right|^{-2\delta_{N,i}}.$$

Using $\varepsilon_N = A_p \left(I_i^N \right)$ this gives

$$\frac{\varepsilon_N}{1 + \eta(\delta_{N,i})} \left| I_i^N \right|^{2\delta_{N,i}} \leq A_{q_N(\cdot)} \left(I_i^N \right) \leq \frac{\varepsilon_N}{1 - \eta(\delta_{N,i})} \left| I_i^N \right|^{-2\delta_{N,i}},$$

and the claim follows.

Claim 4. For all N and all $i \in \{1, 2, ..., N\}$,

$$\left| I_i^N \right|^{-\delta_{N,i}} \leq e^{\psi(|I_i^N|)}.$$

Fix I_i^N. The function p is continuous on I since it belongs to $\mathcal{SLH}(I)$. As

$$p_{I_i^N}^+ - p_{I_i^N}^- = \delta_{N,i},$$

there are points $x, y \in I_i^N$ with $|p(x) - p(y)| = \delta_{N,i}$. Using (5.5) we obtain

$$\left|I_i^N\right|^{-\delta_{N,i}} \leq |x - y|^{-|p(x)-p(y)|} \leq e^{\psi(|x-y|)} \leq e^{\psi\left(\left|I_i^N\right|\right)}.$$

Claim 5. There is a constant $C > 0$ such that for all N and all $i \in \{1, 2, ..., N\}$,

$$C^{-1}\varepsilon_N \leq \left|I_i^N\right| \leq C\varepsilon_N.$$

Since $q_N = p_{I_i^N}^- + \delta_{N,i} := r_{N,i}$ is a constant function on I_i^N, by Lemma 5.11 we have

$$A_{q_N(\cdot)}\left(I_i^N\right) = \mathfrak{B}\left(r_{N,i}, r_{N,i}\right)\left|I_i^N\right|.$$

It is easy to see that there exists $a > 0$ such that $a^{-1} \leq \mathfrak{B}\left(r_{N,i}, r_{N,i}\right) \leq a$ for all N and all $i \in \{1, 2, ..., N\}$. Using Claim 4 we see that

$$\left|I_i^N\right|^{-2\delta_{N,i}} \leq e^{2\psi\left(\left|I_i^N\right|\right)} \leq e^{2\psi(|I|)} := K,$$

and by Claim 3,

$$K^{-1}\beta_N^{-1}\varepsilon_N \leq \mathfrak{B}\left(r_{N,i}, r_{N,i}\right)\left|I_i^N\right| \leq K\beta_N\varepsilon_N.$$

Hence

$$a^{-1}K^{-1}\beta_N^{-1}\varepsilon_N \leq \left|I_i^N\right| \leq aK\beta_N\varepsilon_N.$$

Since $\beta_N \to 1$ as $N \to \infty$, the claim follows.

We can now proceed to finish the proof of the proposition. Since by Claim 1, $\varepsilon_N \to 0$ one has from, Claim 5 that

$$\max\left\{\left|I_i^N\right| : i = 1, ..., N\right\} \to 0 \text{ as } N \to \infty;$$

Claim 4 yields then:

$$\left|I_i^N\right|^{-2\delta_{N,i}} \leq e^{2\psi\left(\left|I_i^N\right|\right)} \leq e^{2\psi\left(\max\left\{\left|I_i^N\right|:i=1,...,N\right\}\right)} := \gamma_N \to 1 \text{ as } N \longrightarrow \infty.$$

Put $\alpha_N = \beta_N\gamma_N$: then

$$\alpha_N \to 1 \text{ as } N \longrightarrow \infty$$

and, by virtue of Claim 2,

$$\alpha_N^{-1}\varepsilon_N \leq A_{q_N(\cdot)}\left(I_i^N\right) \leq \alpha_N\varepsilon_N. \tag{5.6}$$

Moreover, invoking again Claim 5 we have

$$N\varepsilon_N = C\sum_{i=1}^{N} C^{-1}\varepsilon_N \leq C\sum_{i=1}^{N}\left|I_i^N\right| = C\left|I\right|,$$

which gives, by (5.6),

$$N\varepsilon_N \left(\alpha_N^{-1}\varepsilon_N - 1\right) = \sum_{i=1}^{N} \left(\alpha_N^{-1}\varepsilon_N - \varepsilon_N\right) \leq \sum_{i=1}^{N} A_{q_N(\cdot)}\left(I_i^N\right) - N\varepsilon_N$$

$$\leq \sum_{i=1}^{N} \left(\alpha_N\varepsilon_N - \varepsilon_N\right) = N\varepsilon_N \left(\alpha_N - 1\right).$$

Thus

$$\left|\sum_{i=1}^{N} A_{q_N(\cdot)}\left(I_i^N\right) - N\varepsilon_N\right| \to 0 \text{ as } N \to \infty.$$

On the other hand we have, by Lemma 5.11 (recall again that q_N is constant on I_i^N),

$$\sum_{i=1}^{N} A_{q_N(\cdot)}\left(I_i^N\right) = \frac{1}{2\pi} \sum_{i=1}^{N} \left(q_N'(\cdot)q_N(\cdot)^{q_N(\cdot)-1}\right)^{1/q_N(\cdot)} \sin\left(\pi/q_N(\cdot)\right)\left|I_i^N\right|$$

$$\to \frac{1}{2\pi} \int_I \left(p'(x)p(x)^{p(x)-1}\right)^{1/p(x)} \sin(\pi/p(x))dx, \text{ as } N \to \infty$$

from which it is easy to see that

$$\lim_{N \to \infty} N\varepsilon_N = \frac{1}{2\pi} \int_I \left(p'(x)p(x)^{p(x)-1}\right)^{1/p(x)} \sin(\pi/p(x))dx.$$

Since ε_N depends monotonically on N it is not difficult to see that

$$\lim_{N \to \infty} N\varepsilon_N = \lim_{\varepsilon \to 0} \varepsilon N(\varepsilon)$$

and consequently

$$\lim_{\varepsilon \to 0} \varepsilon N(\varepsilon) = \frac{1}{2\pi} \int_I \left(p'(x)p(x)^{p(x)-1}\right)^{1/p(x)} \sin(\pi/p(x))dx.$$

The proof is complete. $\qquad\qquad\qquad\qquad\qquad\qquad\qquad\qquad\qquad\square$

Given $p \in \mathcal{SLH}(I)$, we construct step-functions that are approximations to p. Let $N \in \mathbb{N}$ and use Lemma 5.14, applied to the function $D := D_p$: there exists $\varepsilon > 0$ such that $N_D(\varepsilon) = N$ and there are non-overlapping intervals I_i^D ($i = 1, ..., N$) that cover I and are such that $D(I_i^D) = \varepsilon$ for $i = 1, ..., N$. Define

$$p_{D,N}^+(x) = \sum_{i=1}^{N} p_{I_i^D}^+ \chi_{I_i^D}(x), \quad p_{D,N}^-(x) = \sum_{i=1}^{N} p_{I_i^D}^- \chi_{I_i^D}(x);$$

step-functions $p_{B,N}^+$ and $p_{B,N}^-$ are defined in the obvious similar way, with the function B in place of D and with intervals I_i^B like those used in the portion of Lemma 5.14 related to B.

Lemma 5.19. *Let $p \in \mathcal{P}(I)$ and $N \in \mathbb{N}$. Let $\varepsilon > 0$ correspond to N in the sense of Lemma 5.14, applied to B, so that $N_B(\varepsilon) = N$, and write*
$$p^-(x) = p_{B,N}^-(x), \quad p^+(x) = p_{B,N}^+(x),$$
where $p_{B,N}^+$ and $p_{B,N}^-$ are defined as indicated above. Then
$$a_{N+1}\left(E_0 : \overset{\circ}{W}_{p^-(\cdot)}^1(I) \to L_{p^+(\cdot)}(I) \right) \leq \varepsilon.$$

Proof.   In the notation of Lemma 5.14, there are intervals I_i^B such that $B(I_i^B) = \varepsilon$ for $i = 1, ..., N$. For each i there exists $y_i \in I_i^B$ such that
$$B(I_i^B) = \sup\left\{ \|f\|_{p^+, I_i^B} : \|f'\|_{p^-, I_i^B} \leq 1, f(y_i) = 0 \right\}.$$
Define
$$P_N f(x) = \sum_{i=1}^{N} f(y_i) \cdot \chi_{I_i^B}(x);$$
it is apparent that P_N has rank N. Let $f \in \overset{\circ}{W}_{p^-}^1(I)$ and set
$$\lambda_0 = \varepsilon \|f'\|_{p^-, I}. \tag{5.7}$$
Then
$$1 = \int_I \left| \frac{f'(x)}{\lambda_0/\varepsilon} \right|^{p^-(x)} dx = \sum_{i=1}^{N} \int_{I_i^B} \left| \frac{f'(x)}{\lambda_0/\varepsilon} \right|^{p^-(x)} dx.$$
Recall that on I_i^B the functions p^- and p^+ have constant values p_i^-, p_i^+, say, with $p_i^+/p_i^- \geq 1$. Thus
$$1 \geq \sum_{i=1}^{N} \left(\int_{I_i^B} \left| \frac{f'(x)}{\lambda_0/\varepsilon} \right|^{p_i^-} dx \right)^{p_i^+/p_i^-} = \sum_{i=1}^{N} (\varepsilon/\lambda_0)^{p_i^+} \left(\int_{I_i^B} |f'(x)|^{p_i^-} dx \right)^{p_i^+/p_i^-}.$$
Observing that
$$\varepsilon = \sup_f \left(\int_{I_i^B} |f(x) - f(y_i)|^{p_i^+} dx \right)^{1/p_i^+} \Big/ \left(\int_{I_i^B} |f'(y)|^{p_i^-} dy \right)^{1/p_i^-}$$
one readily gets
$$1 \geq \sum_{i=1}^{N} (1/\lambda_0)^{p_i^+} \int_{I_i^B} |f(x) - f(y_i)|^{p_i^+} dx = \sum_{i=1}^{N} \int_{I_i^B} \left| \frac{f(x) - f(y_i)}{\lambda_0} \right|^{p_i^+} dx$$
$$= \int_I \left| \frac{(E_0 - P_N)(f)(x)}{\lambda_0} \right|^{p^+(x)} dx,$$
from which it follows that $\|(E_0 - P_N)f\|_{p^+, I} \leq \lambda_0$. Using the definition (5.7) of λ_0 we see that
$$\|(E_0 - P_N)f\|_{p^+, I} \leq \varepsilon \|f'\|_{p^-, I},$$
and so $a_{N+1}\left(E_0 : \overset{\circ}{W}_{p^-(\cdot)}^1(I) \to L_{p^+}(I) \right) \leq \varepsilon$, as claimed.   □

We next obtain a lower estimate for the Bernstein numbers.

Lemma 5.20. *Let $p \in \mathcal{P}(I)$ and $N \in \mathbb{N}$. Let $\varepsilon > 0$ correspond to N in the sense of Lemma 5.14, applied to D, so that $N_D(\varepsilon) = N$, and write*

$$p^-(x) = p^-_{D,N}(x), \ \ p^+(x) = p^+_{D,N}(x),$$

where $p^+_{D,N}$ and $p^-_{D,N}$ are defined as indicated above. Then

$$b_N \left(E_0 : \overset{\circ}{W}^1_{p^+}(I) \to L_{p^-}(I) \right) \geq \varepsilon.$$

Proof. In the notation of Lemma 5.14, there are intervals I_i^D such that $D(I_i^D) = \varepsilon$ for $i = 1, ..., N$. Since E_0 is compact, for each i there exists $f_i \in \overset{\circ}{W}^1_{p^+(\cdot)}(I_i^D)$, with f_i' equal to 0 outside I_i^D, such that

$$\|E_0 f_i\|_{p^-, I_i^D} / \|f_i'\|_{p^+, I_i^D} = \varepsilon, \tag{5.8}$$

and $f_i(c_i) = f_i(c_{i+1}) = 0$, where c_i and c_{i+1} are the endpoints of I_i^D. On each I_i^D the functions p^- and p^+ are constant; denote these constant values by p_i^- and p_i^+, respectively and note that $p_i^- / p_i^+ \leq 1$. Set

$$X_N = \left\{ f = \sum_{i=1}^{N} \alpha_i f_i : \alpha_i \in \mathbb{R} \right\}.$$

Thus $\dim X_N = N$. Choose any non-zero $f \in X_N$ and put

$$\lambda_0 = \varepsilon \|f'\|_{p^+, I}.$$

Then

$$
\begin{aligned}
1 &= \int_I \left| \frac{f'(x)}{\lambda_0/\varepsilon} \right|^{p^+(x)} dx = \sum_{i=1}^{N} \int_{I_i^D} \left| \frac{f'(x)}{\lambda_0/\varepsilon} \right|^{p_i^+(x)} dx \\
&\leq \sum_{i=1}^{N} \left(\int_{I_i^D} \left| \frac{f'(x)}{\lambda_0/\varepsilon} \right|^{p_i^+(x)} dx \right)^{p_i^-/p_i^+} \\
&= \sum_{i=1}^{N} (\varepsilon/\lambda_0)^{p_i^-} \left(\int_{I_i^D} |f'(x)|^{p_i^+(x)} dx \right)^{p_i^-/p_i^+} \\
&= \sum_{i=1}^{N} (\varepsilon/\lambda_0)^{p_i^-} \left(\int_{I_i^D} |\alpha_i f_i'(x)|^{p_i^+} dx \right)^{p_i^-/p_i^+}.
\end{aligned}
$$

A direct application of (5.8) now yields:

$$1 \leq \sum_{i=1}^{N} (1/\lambda_0)^{p_i^-} \int_{I_i^D} |E_0(\alpha_i f_i)(x)|^{p_i^-} \, dx$$

$$= \sum_{i=1}^{N} \int_{I_i^D} \left| \frac{f(x)}{\lambda_0} \right|^{p_i^-} \, dx = \int_I \left| \frac{f(x)}{\lambda_0} \right|^{p^-(x)} \, dx,$$

from which it follows that

$$\varepsilon \leq b_N \left(T : \overset{\circ}{W}{}^1_{p^-(\cdot)}(I) \to L_{p^-(\cdot)}(I) \right),$$

and the proof is complete. $\hfill\square$

Theorem 5.10. *Let $p \in \mathcal{P}(I)$ be continuous on I. For all $N \in \mathbb{N}$ let ε_N be defined by the equality $N = N_B(\varepsilon_N)$. Then there are sequences K_N, L_N, with $K_N \to 1$ and $L_N \to 1$ as $N \to \infty$, such that*

(i) $a_{N+1} \left(E_0 : \overset{\circ}{W}{}^1_p(I) \to L_p(I) \right) \leq K_N \varepsilon_N$,

(ii) $b_N \left(E_0 : \overset{\circ}{W}{}^1_p(I) \to L_p(I) \right) \geq L_N \varepsilon_N$.

Proof. Because of the multiplicative property (S3) of the approximation numbers, $a_{N+1} \left(E_0 : \overset{\circ}{W}{}^1_p(I) \to L_p(I) \right)$ is majorised by

$$\left\| id_N^- : \overset{\circ}{W}{}^1_{p)}(I) \to \overset{\circ}{W}{}^1_{p_{B,N}^-}(I) \right\|$$

$$\times a_{N+1}(E_0 : \overset{\circ}{W}{}^1_{p_{B,N}^-}(I) \to L_{p_{B,N}^+}(I)) \times \left\| id_N^+ : L_{p_{B,N}^+(\cdot)}(I) \to L_p(I) \right\|,$$

where id_N^- and id_N^+ are the obvious embedding maps, while $p_{B,N}^+$ and $p_{B,N}^-$ have the same meaning as in Lemma 5.19. Let I_i^B be also as defined in Lemma 5.19. Since $|I_i^B| \to 0$ when $N \to \infty$, and p is continuous, it is clear that

$$\left\| p - p_{B,N}^- \right\|_{\infty,I} \to 0 \text{ and } \left\| p - p_{B,N}^+ \right\|_{\infty,I} \to 0.$$

Thus by Corollary 1.6,

$$\left\| id_N^- : \overset{\circ}{W}{}^1_{p(\cdot)}(I) \to \overset{\circ}{W}{}^1_{p_{B,N}^-}(I) \right\| \to 1 \text{ and } \left\| id_N^+ : L_{p_{B,N}^+}(I) \to L_p(I) \right\| \to 1$$

as $N \to \infty$. Part (i) now follows from Lemma 5.19. The proof of (ii) is similar and requires the use of Lemma 5.20 instead. $\hfill\square$

Theorem 5.11. *Let $p \in \mathcal{SLH}(I)$. Then*

$$\lim_{n \to \infty} n s_n (E_0 : \overset{\circ}{W}^1_{p(\cdot)}(I) \to L_{p(\cdot)}(I))$$

$$= \frac{1}{2\pi} \int_I \left\{ p'(t) p(t)^{p(t)-1} \right\}^{1/p(t)} \sin\left(\pi/p(t) \right) dt,$$

where s_n denotes the nth approximation, Gelfand, Kolmogorov or Bernstein number of E_0.

Proof. Use Theorem 5.10, Proposition 5.1 and the inequalities of Theorems 5.4 and 5.5. □

The proofs of Theorems 5.9 and 5.11 may be combined to yield the following theorem, which contains both these results.

Theorem 5.12. *Let I be representable as the finite union of non-overlapping intervals J_i $(i = 1, ..., m)$ and suppose that $p \in \mathcal{SLH}(I_i)$ for each $i \in \{1, 2, ..., m\}$. Then*

$$\lim_{n \to \infty} n s_n (E_0 : \overset{\circ}{W}^1_p(I) \to L_p(I))$$

$$= \frac{1}{2\pi} \int_I \left\{ p'(t) p(t)^{p(t)-1} \right\}^{1/p(t)} \sin\left(\pi/p(t) \right) dt,$$

where s_n denotes the nth approximation, Gelfand, Kolmogorov or Bernstein number of E_0.

5.3 Integral Operators

On a compact interval $I = [a, b]$ we consider the integral operator

$$T_{v,u} f(x) := v(x) \int_a^x f(t) u(t) dt, \ (x \in I),$$

as a map between $L_p(I)$ spaces. If no ambiguity is likely we shall simply denote this map also by $T_{a,(a,b)}$, T_a or just T. Here u and v are given real variable functions with $|\operatorname{supp} u| = |\operatorname{supp} v| = |I|$.

When p and q are constants and $1 \le p \le q \le \infty$ we require that for all $X \in (a, b)$,

$$u \in L_{p'}(a, X), \text{ and } v \in L_q(X, b);$$

in this setting T is a bounded map from $L_p(I)$ into $L_q(I)$ if and only if

$$A := \sup_{x \in (a,b)} \|u \chi_{(a,x)}\|_{p',(a,b)} \|v \chi_{(x,b)}\|_{q,(a,b)} < \infty.$$

Moreover we have that $A \le \|T|L_p(I) \to L_q(I)\| \le 4A$ (see [28], Theorem 2.2.1).

Let us set:

$$A(c,d) := \sup_{c < X < d} \{\|u\|_{p',(c,X)} \|v\|_{q,(X,d)}\}.$$

When $1 < p \le q < \infty$ and T is a bounded map from $L_p(I)$ into $L_q(I)$ then the condition:

$$\lim_{s \to a_+} A(a,s) = \lim_{s \to b_-} A(s,b) = 0$$

guarantees the compactness of T (see [28], Theorem 2.3.1).

In the case when T is a compact map from $L_p(I)$ into $L_p(I)$ one can estimate the speed of decay of its s-numbers under some additional conditions on weights, as we can see from the next theorem (see [31], Theorem 6.4):

Theorem 5.13. *Let* $1 < p < \infty$, $u \in L_p(I)$ *and* $v \in L_{p'}(I)$. *Then*

$$\lim_{n \to \infty} s_n(T_{v,u})n = \gamma_p \int_I |u(t)v(t)|dt,$$

where $\gamma_p = (p')^{1/p} p^{1/p'} \pi^{-1} \sin(\pi/p)/2$ *and* s_n *stands for any strict s-number.*

With a more stringent restriction on the weights u, v one can get bounds for the second asymptotics, though unfortunately in quite an unappealing form.

Theorem 5.14. *Let* $1 < p < \infty$, $u \in L_{p'}(I), v \in L_p(I)$ *and*

$$(v'/v), (u'/u) \in L_1(I) \cap C(I).$$

Then

$$\limsup_{n \to \infty} \left| n \left[n s_n(T) - \gamma_p \int_I |u(x)v(x)|dx \right] \right|$$

$$\le \int_I |u(x)v(x)|dx \left[\int_I \left| \frac{v'(x)}{v(x)} \right| dx + \int_I \left| \frac{u'(x)}{u(x)} \right| dx + \gamma_p \right.$$

$$\left. + \left(\int_I \left| \frac{u'(x)}{u(x)} \right| dx \right) \left(\int_I \left| \frac{v'(x)}{v(x)} \right| dx \right) \right],$$

where $s_n(T)$ *stands for any strict s-number of the Hardy-type operator T.*

This theorem gives the following information about the second asymptotics for strict s-numbers of T:

$$s_n(T) = \frac{\bar{\gamma}_p}{n} \int_I |u(x)v(x)| dx + O(n^{-2}).$$

More details about these estimates can be found in ([31], Chapter 7).

We next assume that $p \in \mathcal{P}^{\log}(I)$ (i.e. $p(x)$ is is globally Hölder-continuous on I, see Definition 2.2). Under this condition we have a characterization of the boundedness for T (see [54]):

Theorem 5.15. *Let $1 < p_- \leq p(x) \leq q(x) \leq q_- < \infty$ and let $p, q \in \mathcal{P}^{\log}(I)$. Then the operator $T_{v,u}$ is bounded from L_p into L_q if and only if*

$$\sup_{a < X < b} \|v\|_{L_q(X,b)} \|u\|_{L_{p'}(a,X)} < \infty.$$

Under the weaker condition that p, q belong to $\mathcal{P}(I)$ the situation is more complicated, as it was shown in [30]. No simple necessary and sufficient condition for the boundedness of T is known in this case.

Utilizing the techniques from Theorem 1 in [29] it is fairly easy to obtain the next claim from Theorem 5.15:

Theorem 5.16. *Let $1 < p_- \leq p(x) \leq q(x) \leq q_- < \infty$ and let $p, q \in \mathcal{P}^{\log}(I)$. Then the operator $T_{v,u}$ is compact from L_p into L_q if and only if*

$$\lim_{s \to a+} \sup_{a < X < s} \|v\|_{L_q(X,s)} \|u\|_{L_{p'}(a,X)} = 0$$

and

$$\lim_{s \to b-} \sup_{s < X < b} \|v\|_{L_q(X,b)} \|u\|_{L_{p'}(s,X)} = 0.$$

For the remaining part of this section we will suppose that

$$1 < p < \infty$$

in Ω, that $p \in \mathcal{P}^{\log}(I)$ and that

$$u \in L_{p'}(I), \text{ and } v \in L_p(I).$$

According to the previous theorem, the above assumptions guarantee that T is a compact map from L_p into L_p.

Lemma 5.21. *Let $1 < c < d < \infty$, $v \in L_d(I)$, $u \in L_{c'}(I)$ and suppose that $p, q \in \mathcal{P}^{\log}(I)$ are such that $p(x), q(x) \in (c, d)$ for all $x \in I$. Then $T_{v,u}$ maps $L_p(I)$ compactly into $L_q(I)$.*

Proof. By Theorem 1.12, $L_p(I)$ and $L_d(I)$ are continuously embedded in $L_c(I)$ and $L_q(I)$, respectively. By Theorem 5.16, T maps $L_c(I)$ compactly into $L_d(I)$. The result now follows by composition of these maps. □

Now we introduce functions which play a paramount role in the determination of the asymptotic behaviour of various s-numbers of T.

Definition 5.10. Let $p, q \in \mathcal{P}^{\log}(I)$, suppose that $J = (c, d) \subset I$, $u \in L_{p'}(J)$, $v \in L_q(J)$, and let $\varepsilon > 0$; set

$$p_J^- = \inf\{p(x) : x \in J\}, \ p_J^+ = \sup\{p(x) : x \in J\}.$$

Then

$$A_{p,q}^{T_{v,u}}(J) := \inf_{y \in J} \sup \left\{ \left\| v(.) \int_y^{\cdot} f(x)u(x)dx \right\|_{q,J} : \|f\|_{p,J} \le 1 \right\},$$

$$B_p^{T_{v,u}}(J) := \inf_{y \in J} \sup \left\{ \left\| v(\cdot) \int_y^{\cdot} f(x)u(x)dx \right\|_{p_J^+,J} : \|f\|_{p_J^-,J} \le 1 \right\},$$

$$C_{p,q}^{T_{v,u}}(J) := \sup \left\{ \|T_{v,u}f\|_{q,J} : \|f\|_{p,J} \le 1, (Tf)(c) = (Tf)(d) = 0 \right\}$$

and

$$D_p^{T_{v,u}}(J) := \sup \left\{ \|T_{v,u}f\|_{p_J^-,J} : \|f\|_{p_J^+,J} \le 1, (Tf)(c) = (Tf)(d) = 0 \right\}.$$

These quantities will sometimes be denoted by $A_{p,q}(c, d)$, etc. Corresponding to these functions we define $N_{A_{p,q}}(\varepsilon)$ to be the minimum of all those $n \in \mathbb{N}$ such that I can be written as $I = \bigcup_{j=1}^n I_j$, where each I_j is a closed sub-interval of I,

$$|I_i \cap I_j| = 0$$

$(i \ne j)$ and $A_{p,q}(I_j) \le \varepsilon$ for every j. The quantities $N_{B_p}(\varepsilon), N_{C_{p,q}}(\varepsilon)$ and $N_{D_p}(\varepsilon)$ are defined in an exactly similar way. For brevity we shall drop superscript $T_{v,u}$ and write

$$A_p(J) = A_{p,p}(J)$$

and $C_p(J) = C_{p,p}(J)$, denoting these quantities by $A_p(J), C_p(J)$ respectively when p is a constant function. When p and q are constant functions we also write

$$A_{p,q}(J) = A_{p,q}(J)$$

and

$$C_{p,q}(J) = C_{p,q}(J).$$

As in Lemma 5.11 we can show that for constant p and q and constant u and v we have the following result:

Lemma 5.22. *Let* $J = (c,d) \subset I$ *and* $p,q \in (1,\infty)$. *Then*

$$A_{p,q}^{T_{v,u}}(J) = C_{p,q}^{T_{v,u}}(J) = \frac{(p'+q)^{1/p-1/q}(p')^{1/q}q^{1/p'}}{2B(1/p',1/q)}|v||u|\,|J|^{1/p'+1/q}$$

$$:= \mathfrak{B}(p,q)|v||u|\,|J|^{1/p'+1/q}. \qquad (5.9)$$

The next corollary is an immediate consequence of the above lemma:

Corollary 5.1. *Let* $p,q \in \mathcal{P}^{\log}(I)$, $u \in L^{p'}(I)$, $v \in L^p(I)$ *and suppose that* $(c,d) \subset I$. *Then, there exist constants* $u_c, v_c > 0$ *where*

$$\operatorname{ess\,inf}_{(c,d)} u \leq u_c \leq \operatorname{ess\,inf}_{(c,d)} u$$

and

$$\inf_{(c,d)} v \leq v_c \leq \sup_{(c,d)} v$$

such that

$$A_{p,q}^{T_{v,u}}(c,d) = A_{p,q}^{T_{v_c,u_c}}(c,d) = \mathfrak{B}(p,q)|v_c||u_c|(d-c)^{1/p'+1/q}.$$

The same holds when A is replaced by any of the quantities B, C or D.

We now set about the task of establishing properties of the quantities introduced in Definition 5.10.

Lemma 5.23. *Let* $p,q \in \mathcal{P}^{\log}(I)$ *and suppose that* $(c,d) \subset I$. *Set*

$$e_- := \min\{p_-, q_-\}, \quad e_+ := \max\{p_+, q_+\}$$

and suppose that $v \in L_{e_+}(c,d)$, $u \in L_{e'_-}(c,d)$.

Then the functions $A_{p,q}^{T_{v,u}}(c,t)$, $B_p^{T_{v,u}}(c,t)$, $C_{p,q}^{T_{v,u}}(c,t)$ *and* $D_p^{T_{v,u}}(c,t)$ *of the variable t are non-decreasing and continuous; so are* $A_{p,q}^{T_{v,u}}(t,d)$, $B_p^{T_{v,u}}(t,d)$, $C_{p,q}^{T_{v,u}}(t,d)$ *and* $D_p^{T_{v,u}}(t,d)$.

Proof. We start with $A := A_{p,q}^{T_{v,u}}$ and prove first that for $h \geq 0$, one has

$$A(c,d) \leq A(c,d+h).$$

Clearly

$$A(c,d+h) = \inf_{y \in (c,d+h)} \sup \left\{ \left\| v(\cdot) \int_y^{\cdot} f(t)u(t)dt \right\|_{q,(c,d+h)} : \|f\|_{p,(c,d+h)} \leq 1 \right\}$$

$$= \min\{X, Y\},$$

where

$$X = \inf_{y \in (c,d)} \sup \left\{ \left\| v(\cdot) \int_y^{\cdot} f(t)u(t)dt \right\|_{q,(c,d+h)} : \|f\|_{p,(c,d+h)} \leq 1 \right\}$$

and

$$Y = \inf_{y \in (d,d+h)} \sup \left\{ \left\| v(\cdot) \int_y^{\cdot} f(t)u(t)dt \right\|_{q,(c,d+h)} : \|f\|_{p,(c,d+h)} \leq 1 \right\}.$$

Now

$$X \geq \inf_{y \in (c,d)} \sup \left\{ \left\| v(\cdot) \int_y^{\cdot} f(t)u(t)dt \right\|_{q,(c,d)} : \|f\|_{p,(c,d)} \leq 1 \right\} = A(c,d)$$

and

$$Y \geq \inf_{y \in (d,d+h)} \sup \left\{ \left\| v(\cdot) \int_y^{\cdot} f(t)u(t)dt \right\|_{q,(c,d)} : \|f\|_{p,(c,d)} \leq 1 \right\}$$

$$\geq \sup \left\{ \left\| v(\cdot) \int_d^{\cdot} f(t)u(t)dt \right\|_{q,(c,d)} : \|f\|_{p,(c,d)} \leq 1 \right\}$$

$$\geq \inf_{y \in (c,d)} \sup \left\{ \left\| v(\cdot) \int_y^{\cdot} f(t)u(t)dt \right\|_{q,(c,d)} : \|f\|_{p,(c,d)} \leq 1 \right\} = A(c,d),$$

which gives $A(c, d + h) \geq A(c, d)$. We next tackle the continuity of A. To this end, suppose that $y \in (c, d+h)$. Then by Hölder's inequality (Theorem 1.5) we have that for some $\alpha \geq 1$ (independent of f, c, d, h, x and y),

$$\left\| v(x) \int_y^x f(t)u(t)dt \right\|_{q,(d,d+h)} \leq \alpha \|v\|_{q,(d,d+h)} \|u\|_{p',(c,d+h)} \|f\|_{p,(c,d+h)}.$$

With the understanding that, unless otherwise specified, the suprema are taken over all f with $\|f\|_{p,(c,d+h)} \leq 1$, we have

$$
\begin{aligned}
A(c,d+h) &\leq \inf_{y \in (c,d+h)} \sup \left\{ \left\| v(\cdot) \int_y^{\cdot} f(t)u(t)dt \right\|_{q,(c,d)} \right. \\
&\qquad\qquad\qquad \left. + \left\| v(\cdot) \int_y^{\cdot} f(t)u(t)dt \right\|_{q,(d,d+h)} \right\} \\
&\leq \inf_{y \in (c,d+h)} \sup \left\{ \left\| v(\cdot) \int_y^{\cdot} f(t)u(t)dt \right\|_{q,(c,d)} \right. \\
&\qquad\qquad\qquad \left. + \alpha \|v\|_{q,(d,d+h)} \|u\|_{p',(c,d+h)} \|f\|_{p',(c,d+h)} \right\} \\
&\leq \inf_{y \in (c,d+h)} \sup \left\| v(\cdot) \int_y^{\cdot} f(t)u(t)dt \right\|_{q,(c,d)} \\
&\qquad\qquad\qquad + \alpha \|u\|_{p',(c,d+h)} \|v\|_{q,(d,d+h)} \\
&\leq \inf_{y \in (c,d)} \sup \left\{ \left\| v(\cdot) \int_y^{\cdot} f(t)u(t)dt \right\|_{q,(c,d)} : \|f\|_{p,(c,d)} \leq 1 \right\} \\
&\qquad\qquad\qquad + \alpha \|u\|_{p',(c,d+h)} \|v\|_{q,(d,d+h)} \\
&= A(c,d) + \alpha \|u\|_{p',(c,d+h)} \|v\|_{q,(d,d+h)}.
\end{aligned}
$$

Since $q \in \mathcal{P}(I)$ and $v \in L_q(d, d+h)$ we know that $\|v\|_{q,(d,d+h)} \to 0$ as $h \to 0$, and so $A(c, \cdot)$ is right-continuous. Left-continuity is proved analogously and the continuity of $A(c, \cdot)$ follows. The arguments for B, C and D are similar and will be omitted. $\qquad\qquad\square$

As an immediate consequence of the result just proved and of Lemma 5.21 we have:

Lemma 5.24. *Let $p \in \mathcal{P}^{\log}(I)$, $\delta > 0$ and suppose that $v \in L_{p+\delta}(I)$, $u \in L_{p'+\delta}(I)$. Then $T : L_p(I) \to L_p(I)$ is compact and for all small enough $\varepsilon > 0$ (depending on δ and p) the quantities $N_{A_p}(\varepsilon)$, $N_{B_p}(\varepsilon)$, $N_{C_{p,q}}(\varepsilon)$ and $N_{D_p}(\varepsilon)$ are well defined and finite.*

We also have:

Lemma 5.25. *Let $p \in \mathcal{P}^{\log}(I)$, $\delta > 0$ and suppose that $v \in L_{p+\delta}(I)$, $u \in L_{p'+\delta}(I)$. Write $A = A_p^{T_{v,u}}$. Then for any large enough (depending on δ and p) $N \in \mathbb{N}$, there exists a unique $\varepsilon > 0$ such that $N_A(\varepsilon) = N$, and there is a covering of I by non-overlapping intervals I_A^i ($i = 1, ..., N$) such that $A(I_A^i) = \varepsilon$ for $i = 1, ..., N$. The same holds when A is replaced by B, C, D.*

Proof. Existence follows from the continuity properties established in Lemma 5.23. For uniqueness, observe that given any two such coverings of I, $\{I_A^i\}_{i=1}^N$ and $\{J_A^i\}_{i=1}^N$, there are m, j, k, l such that $I_A^m \subset J_A^j$ and $J_A^k \subset I_A^l$. Assuming that $A(I_A^i) = \varepsilon_1$ and $A(J_A^i) = \varepsilon_2$, we obtain $\varepsilon_1 \leq \varepsilon_2 \leq \varepsilon_1$ by the monotonicity of A. $\square$

5.3.1   *The case when p is a step-function*

Let $\{J_i\}_{i=1}^m$ be a covering of I by non-overlapping intervals and let p be the step-function defined by

$$p(x) = \sum_{i=1}^m \chi_{J_i}(x) p_i, \qquad (5.10)$$

where $p_i \in (1, \infty)$ for each i. In this section we suppose $v \in L_p(I)$, $u \in L_{p'}(I)$ and we shall write $A = A_p^{T_{v,u}}$ as usual, B, C, D will have analogous meaning.

Lemma 5.26. *Let p be the step-function given by (5.10). Then*

$$T_{v,u} : L_p(I) \to L_p(I)$$

is compact and for small enough $\varepsilon > 0$,
    (i) $b_{N_C(\varepsilon)-m}(T) > \varepsilon$ and
    (ii) $a_{N_A(\varepsilon)+2m-1}(T) < \varepsilon$.

Proof. From Theorem 5.16 we can quite simply obtain the compactness of $T_{v,u}$ on each J_i and since m is finite we have compactness of $T_{v,u}$ on I. Let $\varepsilon > 0$: Then compactness of $T_{v,u}$ guarantees the finiteness of $N_A(\varepsilon)$ and $N_C(\varepsilon)$.

(i) Let $c \in I_i$. Then from Lemma 5.23 we have the continuity of $C(c, \cdot)$ on I_i. For given $\varepsilon > 0$ there is a set $\{I_i : i = 1, ..., N_C(\varepsilon)\}$ of non-overlapping intervals covering I, such that $C(I_i) = \varepsilon$ whenever $1 \leq i < N_C(\varepsilon)$ and $C\left(I_{N_C(\varepsilon)}\right) \leq \varepsilon$.

Let $\eta \in (0, \varepsilon)$. Then corresponding to each i with $1 \leq i < N_C(\varepsilon)$, there is a function f_i such that supp $f_i \subset I_i := (a_i, a_{i+1})$,

$$\|f_i\|_p = 1,$$

$$\varepsilon - \eta < \|T_{v,u} f_i\|_p \leq \varepsilon$$

and

$$(T_{v,u} f)(a_i) = (T_{v,u} f)(a_{i+1}) = 0.$$

Let $\{I_{i_k}\}_{k=1}^M$ denote the set of those intervals I_i, $1 \le i < N_C(\varepsilon)$, each of which is contained in one of the intervals J_l from the formula (5.4) of p. Then

$$N_C(\varepsilon) - m \le M \le N_C(\varepsilon).$$

Put

$$X_M = \left\{ f = \sum_{r=1}^M \alpha_{i_r} f_{i_r} : \alpha_{i_r} \in \mathbb{R} \right\};$$

this is an M-dimensional subspace of $L_p(I)$. Note that since p is constant on I_{i_r}, $p(x) = p_{i_r}$ on I_{i_r}. Choose $f \in X_M \backslash \{0\}$. With $\lambda_0 := \|T_{v,u}f\|_{p(\cdot)}$ we have

$$1 \ge \int_I \left| \frac{T_{v,u}f(x)}{\lambda_0} \right|^{p(x)} dx \ge \sum_{r=1}^M \int_{I_{i_r}} \left| \frac{T_{v,u}f(x)}{\lambda_0} \right|^{p(x)} dx$$

$$= \sum_{r=1}^M \left(\frac{1}{\lambda_0} \right)^{p_{i_r}} \int_{I_{i_r}} |T_{v,u}f(x)|^{p_{i_r}} dx \ge \sum_{r=1}^M \left(\frac{\varepsilon - \eta}{\lambda_0} \right)^{p_{i_r}} \int_{I_{i_r}} |f(x)|^{p_{i_r}} dx$$

$$= \sum_{r=1}^M \int_{I_{i_r}} \left| \frac{f(x)}{\lambda_0/(\varepsilon - \eta)} \right|^{p(x)} dx = \int_{\cup_{r=1}^M I_{i_r}} \left| \frac{f(x)}{\lambda_0/(\varepsilon - \eta)} \right|^{p(x)} dx$$

$$= \int_I \left| \frac{f(x)}{\lambda_0/(\varepsilon - \eta)} \right|^{p(x)} dx.$$

Hence

$$\|f\|_{p,I} \le \|T_{v,u}f\|_{p,I} / (\varepsilon - \eta),$$

and so $b_{N_C(\varepsilon)-m}(T_{v,u}) \ge b_M(T_{v,u}) \ge \varepsilon - \eta$.

(ii) This follows along the same lines as (i). In this case, we let $\{I_i\}_{i=1}^{N_A(\varepsilon)}$ be a set of non-overlapping intervals covering I for which

$$A(I_i) = \varepsilon \quad \text{when } i = 1, ..., N_A(\varepsilon) - 1$$

and

$$A\left(I_{N_A(\varepsilon)}\right) \le \varepsilon.$$

Set $\{I_i^+\}_{i=1}^M$ to denote the family of all non-empty intervals for which there exist j and k such that $I_i^+ = I_j \cap I_k$. Clearly $N_A(\varepsilon) \le M \le N_A(\varepsilon) + 2(m-1)$. Let $\eta > 0$. Then given any $i \in \{1, ..., M\}$, there exists $y_i \in I_i^+$ such that

$$\sup \left\{ \left\| v(\cdot) \int_{y_i}^{\cdot} f(t)u(t)dt \right\|_{p,I_i^+} : \|f\|_{p,I_i^+} = 1 \right\} \le \varepsilon + \eta.$$

Define

$$P_\varepsilon(f) = \sum_{i=1}^{M} \left(v(\cdot) \int_a^{y_i} f(t)u(t)dt \right) \chi_{I_i^+}.$$

It is clear that P_ε is a linear map from $L_p(I)$ to $L_p(I)$ with rank M. Let p_i be the constant value of p on I_i^+. Then for any $\lambda_0 \in (0, \infty)$ and $f \in L_p(I)$,

$$\int_I \left| \frac{(T_{v,u} - P_\varepsilon)f(x)}{\lambda_0} \right|^{p(x)} dx = \sum_{i=1}^{M} \int_{I_i^+} \left| \frac{v(x) \int_{y_i}^x f(t)u(t)dt}{\lambda_0} \right|^{p(x)} dx$$

$$= \sum_{i=1}^{M} \lambda_0^{-p_i} \int_{I_i^+} \left| v(x) \int_{y_i}^x f(t)u(t)dt \right|^{p_i} dx$$

$$\leq \sum_{i=1}^{M} \frac{(\varepsilon + \eta)^{p_i}}{\lambda_0^{p_i}} \int_{I_i^+} |f|^{p_i} dx \int_{I_i^+} |v|^{p_i} dx \left(\int_{I_i^+} |u|^{p_i'} dx \right)^{p_i/p_i'}$$

$$\leq \sum_{i=1}^{M} \frac{(\varepsilon + \eta)^{p_i}}{\lambda_0^{p_i}} \left(\int_{I_i^+} |f|^{p_i} dx \right) \rho_{p,I}(v) \, \|u\chi_{I_i}\|_{p',I}^{p_i-1}$$

$$= \left(\int_I \left| \frac{f(x)}{\lambda_0/(\varepsilon + \eta)} \right|^{p(x)} dx \right) \rho_{p,I}(v) \, \max\{1, \|u\|_{p',I}^{p_+-1}\}.$$

Now for $\lambda_0 = (1 - \eta) \|(T_{v,u} - P_\varepsilon)f\|_{p,I}$, one has

$$1 < \int_I \left| \frac{(T - P_\varepsilon)f(x)}{\lambda_0} \right|^{p(x)} dx$$

$$\leq \left(\int_I \left| \frac{f(x)}{\lambda_0/(\varepsilon + \eta)} \right|^{p(x)} dx \right) \rho_{p,I}(v) \, \max\{1, \|u\|_{p',I}^{p_+-1}\},$$

from which we see that

$$\|f\|_{p,I} \, \rho_{p,I}(v) \, \max\{1, \|u\|_{p',I}^{p_+-1}\} > (1 - \eta) \|(T_{v,u} - P_\varepsilon)f\|_{p,I} / (\varepsilon + \eta),$$

so that

$$\max\{1, \|u\|_{p',I}^{p_+-1}\} \, \rho_{p,I}(v) \frac{\varepsilon + \eta}{1 - \eta} > \frac{\|(T - P_\varepsilon)f\|_{p,I}}{\|f\|_{p,I}}.$$

The proof is concluded by letting $\eta \to 0$. $\qquad\square$

Lemma 5.27. *Let p be the step-function given by (5.4) and $v \in L_p(I)$, $u \in L_{p'}(I)$. Then*

$$\lim_{\varepsilon \to 0} \varepsilon N(\varepsilon) = \frac{1}{2\pi} \int_I |v||u| \left(p'(x)p(x)^{p(x)-1} \right)^{1/p(x)} \sin(\pi/p(x))dx,$$

where N stands for any of the quantities N_A, N_B, N_C or N_D.

Proof. Use the fact that p is a step function together with Lemmas 5.22 and 5.25. □

The main result for a step-function exponent p is given below.

Theorem 5.17. *Let p be the step-function given by (5.10) and $v \in L_p(I)$, $u \in L_{p'}(I)$. Then for the compact map*

$$T_{v,u} : L_p(I) \to L_p(I)$$

we have

$$\lim_{n \to \infty} n s_n(T_{v,u}) = \frac{1}{2\pi} \int_I |u||v| \left(p'(x) p(x)^{p(x)-1} \right)^{1/p(x)} \sin(\pi/p(x)) dx,$$

where $s_n(T)$ denotes the nth approximation, Gelfand, Kolmogorov or Bernstein number of T.

Proof. From Lemma 5.26 together with Theorems 5.4 and 5.5 it follows that

$$\varepsilon N_A(\varepsilon) \geq a_{N_A(\varepsilon)+2m-1}(T_{v,u})N_A(\varepsilon) \geq b_{N_A(\varepsilon)+2m-1}(T_{v,u})N_A(\varepsilon)$$

and that

$$\varepsilon N_C(\varepsilon) \leq b_{N_C(\varepsilon)-m}(T_{v,u})N_C(\varepsilon).$$

The result for the approximation and Bernstein numbers is obtained by applying Lemma 5.27. The rest follows from the inequalities of Theorems 5.4 and 5.5. □

5.3.2 *The case of strongly log-Hölder-continuous p*

In order to obtain a result in this case that is similar to that of Theorem 5.17, we approximate p by step-functions. Some technical lemmas, which we now present, will pave the ground for this task.

Lemma 5.28. *Let $\delta > 0$, let $J \subset I$ be an interval and suppose that $p, q \in \mathcal{P}(J)$ satisfy the conditions:*

$$p(x) \leq q(x) \leq p(x) + \delta \text{ for all } x \in J.$$

Then

$$\left(\delta |J| + \delta^{-\delta} \right)^{-2} A_{p+\delta,p}^{T_{v,u}}(J) \leq A_q^{T_{v,u}}(J) \leq \left(\delta |J| + \delta^{-\delta} \right)^2 A_{p(\cdot),p+\delta}(J).$$

Proof. Set

$$B_1 = \{f : \|f\|_q \le 1\}, B_2 = \{f : \|f\|_p \le \delta\,|J| + \delta^{-\delta}\},$$

where the norms are with respect to the interval J. By Theorem 1.13, one has

$$\|f\|_p \le \left(\delta\,|J| + \delta^{-\delta}\right)\|f\|_q,$$

which gives $B_1 \subset B_2$ and

$$
\begin{aligned}
A_q^{T_{v,u}}(J) &= \inf_{y \in J} \sup \left\{ \left\| v(\cdot) \int_y^{\cdot} f(t)u(t)dt \right\|_q : f \in B_1 \right\} \\
&\le \inf_{y \in J} \sup \left\{ \left(\delta\,|J| + \delta^{-\delta}\right)\left\| v(\cdot) \int_y^{\cdot} f(t)u(t)dt \right\|_{p+\delta} : f \in B_2 \right\} \\
&= \left(\delta\,|J| + \delta^{-\delta}\right)^2 \inf_{y \in J} \sup \left\{ \left\| \int_y^{\cdot} \frac{f(t)u(t)}{\delta\,|J| + \delta^{-\delta}}dt \right\|_{p+\delta} : \left\| \frac{f}{\delta\,|J| + \delta^{-\delta}} \right\|_p \le 1 \right\} \\
&= \left(\delta\,|J| + \delta^{-\delta}\right)^2 \inf_{y \in J} \sup \left\{ \left\| v(\cdot) \int_y^{\cdot} g(t)u(t)dt \right\|_{p+\delta} : \|g\|_p \le 1 \right\} \\
&= \left(\delta\,|J| + \delta^{-2}\right)^{-2} A_{p,p+\delta}^{T_{v,u}}(J).
\end{aligned}
$$

The proof of the remaining part of the claimed inequality is similar. $\qquad\square$

Lemma 5.29. *Let $J \subset I$ be an interval with $|J| \le 1$. Suppose that $\widetilde{p} \in (1, \infty)$ that*

$$\inf_J |u|, \inf_J |v| > 0$$

and that

$$\sup_J |u|, \sup_J |v| < \infty.$$

Then there is a bounded positive function η defined on $(0,1)$, with $\eta(\delta) \to 0$ as $\delta \to 0$, such that if $p, q \in \mathcal{P}(J)$ with

$$\widetilde{p} \le p(x) \le \widetilde{p} + \delta, \ \widetilde{p} \le q(x) \le \widetilde{p} + \delta \ in \ J,$$

then

$$(1-\eta(\delta))\,|J|^{2\delta}\,\frac{\inf_J |u| \inf_J |v|}{\sup_J |u| \sup_J |v|} \le \frac{A_p(J)}{A_q(J)} \le (1+\eta(\delta))\,|J|^{-2\delta}\,\frac{\sup_J |u| \sup_J |v|}{\inf_J |u| \inf_J |v|}.$$

Proof. We prove only the right-hand side inequality as the rest follows in a similar fashion. By Lemma 5.22 and (5.9) we have

$$\frac{A_p(J)}{A_q(J)} \leq \left(\delta |J| + \delta^{-\delta}\right)^4 \frac{A_{\widetilde{p},\widetilde{p}+\delta}(J)}{A_{\widetilde{p}+\delta,\widetilde{p}}(J)}$$

$$\leq \left(\delta |J| + \delta^{-\delta}\right)^4 \frac{\mathfrak{B}(\widetilde{p},\widetilde{p}+\delta) \, \sup_J |u| \sup_J |v|}{\mathfrak{B}(\widetilde{p}+\delta,\widetilde{p}) \, \inf_J |u| \inf_J |v|} |J|^{-2\delta/(\widetilde{p}(\widetilde{p}+\delta))}$$

$$\leq \left(\delta |J| + \delta^{-\delta}\right)^4 \frac{\mathfrak{B}(\widetilde{p},\widetilde{p}+\delta) \, \sup_J |u| \sup_J |v|}{\mathfrak{B}(\widetilde{p}+\delta,\widetilde{p}) \, \inf_J |u| \inf_J |v|} |J|^{-2\delta}.$$

Since

$$\lim_{\delta \to 0} \left(\delta |J| + \delta^{-\delta}\right)^4 \frac{\mathfrak{B}(\widetilde{p},\widetilde{p}+\delta)}{\mathfrak{B}(\widetilde{p}+\delta,\widetilde{p})} = 1,$$

the choice

$$\eta(\delta) = \max \left\{ \delta, \left(\delta |J| + \delta^{-\delta}\right)^4 \frac{\mathfrak{B}(\widetilde{p},\widetilde{p}+\delta)}{\mathfrak{B}(\widetilde{p}+\delta,\widetilde{p})} - 1 \right\}$$

establishes the lemma. $\qquad\square$

Lemma 5.30. *Let $p \in \mathcal{P}(I), \delta > 0, a_1 < b_1 \leq a_2 < b_2$ and $J_i = (a_i, b_i) \subset I$ $(i = 1, 2)$; let f_1, f_2 be functions on I such that $\operatorname{supp} f_i \subset J_i$ $(i = 1, 2)$ and $\|T_{v,u} f_1\|_{p,J_1} > \delta$. Then*

$$\|T_{v,u}(f_1 - f_2)\|_{p,I} > \delta.$$

Proof. Since $\|T_{v,u}(f_1/\delta)\|_{p,J_1} > 1$ we have

$$\int_{a_1}^{b_1} \left| v(x) \int_{a_1}^{x} \frac{f_1(t)u(t)}{\delta} dt \right|^{p(x)} dx > 1.$$

Thus

$$\int_a^b \left| \frac{T_{v,u}(f_1 - f_2)(x)}{\delta} \right|^{p(x)} dx = \int_a^b \left| v(x) \int_a^x \frac{f_1(t)u(t) - f_2(t)u(t)}{\delta} dt \right|^{p(x)} dx$$

$$\geq \int_{a_1}^{b_1} \left| v(x) \int_a^x \frac{f_1(t)u(t) - f_2(t)u(t)}{\delta} dt \right|^{p(x)} dx$$

$$= \int_{a_1}^{b_1} \left| v(x) \int_a^x \frac{f_1(t)u(t)}{\delta} dt \right|^{p(x)} dx > 1,$$

and so $\|T_{v,u}(f_1 - f_2)\|_{p,I} > \delta$. $\qquad\square$

In what follows we shall require $p \in \mathcal{SLH}(I))$ (see Definition 5.9). We remind the reader that $[a, b]$ is a compact interval.

Proposition 5.2. *Let* $p \in \mathcal{SLH}(I)$, *and* v, u *be continuous functions on* I. *Then*

$$\lim_{\varepsilon \to 0} \varepsilon N(\varepsilon) = \frac{1}{2\pi} \int_I |v(x)||u(x)| \left(p'(x)p(x)^{p(x)-1} \right)^{1/p(x)} \sin\left(\pi/p(x)\right) dx,$$

where N *stands for* $N_{A_p^{T_v,u}}$ *or* $N_{C_p^{T_v,u}}$.

Proof. We prove only the case $N = N_{A_p}$, the other case following in a similar manner. Let $N \in \mathbb{N}$. By Lemma 5.25, there are a constant $\varepsilon_N > 0$ and a set of non-overlapping intervals $\{I_i^N\}_{i=1}^N$ covering I such that $A_p^{T_v,u}(I_i^N) = \varepsilon_N$ for every i. Let us denote by u_N and v_N step functions which are constant on each I_i^N and for which we have $A_p^{T_{v_N},u_N}(I_i^N) = \varepsilon_N$ (from Corollary 5.1 it is clear how to define u_N, v_N on I_i^N).

Let q_N be the step-function defined by

$$q_N(x) = \sum_{i=1}^N p_{I_i^N}^+ \chi_{I_i^N}(x)$$

and set

$$\delta_{N,i} = p_{I_i^N}^+ - p_{I_i^N}^-.$$

Then

$$p(x) \le q_N(x) \le p(x) + \delta_{N,i} \text{ for } i = 1, ..., N.$$

Claim 1. $\varepsilon_N \to 0$ as $N \to \infty$.

To prove this, note that clearly ε_N is non-increasing. Suppose that there exists $\delta > 0$ such that $\varepsilon_N > \delta$ for all N. Fix N and set $I_i^N = I_i = (a_i, a_{i+1})$. Since $A_{p,I_i} > \delta$, for each $i \in \{1, ..., N\}$ there is a function f_i, with supp $f_i \subset I_i$, such that $\|f_i\|_{p,I_i} \le 1$ and $\left\| \int_{a_i}^{\cdot} f_i \right\|_{p,I_i} = \|Tf_i\|_{p,I_i} > \delta$. By Lemma 5.30,

$$\|T(f_i - f_j)\|_{p,I_i} > \delta \text{ for } i < j,$$

and so there are N functions $f_1, ..., f_N$ in the unit ball of $L_{p,I}$ such that

$$\|T(f_i - f_j)\|_{p,I_i} > \delta \text{ for } i \ne j.$$

Since N can be arbitrarily large, this contradicts the compactness of T and establishes the claim.

Claim 2. $\lim_{N \to \infty} \max \left\{ |I_i^N| : i = 1, 2, ..., N \right\} = 0$.

If this were false, there would be sequences $\{N_k, i_k\}_{k=1}^{\infty}$, $i_k \in \{1, 2, ..., N_k\}$, and an interval J such that $J \subset I_{i_k}^{N_k}$ for each k, so that

$$\varepsilon_{N_k} = A_p^{T_{v,u}}\left(I_{i_k}^{N_k}\right) \geq A_p^{T_{v,u}}\left(J\right) > 0,$$

contradicting the fact that $\varepsilon_N \to 0$.

Claim 3. We start by showing that there is a sequence $\{\beta_N\}$, with $\beta_N \downarrow 1$, such that for all $i \in \{1, ..., N\}$,

$$\beta_N^{-1}\varepsilon_N \left|I_i^N\right|^{2\delta_{N,i}} \leq A_{q_N(\cdot)}^{T_{v_N, u_N}}\left(I_i^N\right) \leq \beta_N \varepsilon_N \left|I_i^N\right|^{-2\delta_{N,i}}.$$

This can be established by noting that since $p^- \leq q_N(x)$, $p(x) \leq p^- + \delta_{N,i}$ on I_i^N, Lemma 5.29 implies:

$$(1 - \eta\left(\delta_{N,i}\right))\left|I_i^N\right|^{2\delta_{N,i}} \leq \frac{A_p^{T_{v_N, u_N}}\left(I_i^N\right)}{A_{q_N(\cdot)}^{T_{v_N, u_N}}\left(I_i^N\right)} \leq (1 + \eta\left(\delta_{N,i}\right))\left|I_i^N\right|^{-2\delta_{N,i}}.$$

Using $\varepsilon_N = A_p^{T_{v_N, u_N}}\left(I_i^N\right)$ this gives

$$\frac{\varepsilon_N}{1 + \eta\left(\delta_{N,i}\right)}\left|I_i^N\right|^{2\delta_{N,i}} \leq A_{q_N(\cdot)}^{T_{v_N, u_N}}\left(I_i^N\right) \leq \frac{\varepsilon_N}{1 - \eta\left(\delta_{N,i}\right)}\left|I_i^N\right|^{-2\delta_{N,i}},$$

and the claim follows.

Claim 4. For all N and all $i \in \{1, 2, ..., N\}$,

$$\left|I_i^N\right|^{-\delta_{N,i}} \leq e^{\psi(|I_i^N|)}.$$

Fix I_i^N. The function p is continuous on I since it belongs to $\mathcal{SLH}(I)$. As $p_{I_i^N}^+ - p_{I_i^N}^- = \delta_{N,i}$, there are points $x, y \in I_i^N$ with $|p(x) - p(y)| = \delta_{N,i}$. Using (5.5) we obtain

$$\left|I_i^N\right|^{-\delta_{N,i}} \leq |x - y|^{-|p(x)-p(y)|} \leq e^{\psi(|x-y|)} \leq e^{\psi(|I_i^N|)}.$$

Claim 5. There is a constant $C > 0$ such that for all N and all $i \in \{1, 2, ..., N\}$,

$$C^{-1}\varepsilon_N \leq \left|I_i^N\right| \cdot \left|v_N\chi_{I_i^N}\right| \cdot \left|u_N\chi_{I_i^N}\right| \leq C\varepsilon_N.$$

Since $q_N = p_{I_i^N}^- + \delta_{N,i} := r_{N,i}$ is a constant function on I_i^N, by Lemma 5.11 we have

$$A_{q_N}^{T_{v_N, u_N}}\left(I_i^N\right) = \mathfrak{B}\left(r_{N,i}, r_{N,i}\right)\left|I_i^N\right| \cdot \left|v_N\chi_{I_i^N}\right| \cdot \left|u_N\chi_{I_i^N}\right|.$$

It is easy to see that there exists $a > 0$ such that $a^{-1} \leq \mathfrak{B}\left(r_{N,i}, r_{N,i}\right) \leq a$ for all N and all $i \in \{1, 2, ..., N\}$. Using Claim 4 we see that

$$\left|I_i^N\right|^{-2\delta_{N,i}} \leq e^{2\psi(|I_i^N|)} \leq e^{2\psi(|I|)} := K,$$

and by Claim 3,

$$K^{-1}\beta_N^{-1}\varepsilon_N \leq \mathfrak{B}\left(r_{N,i}, r_{N,i}\right)\left|I_i^N\right| \cdot \left|v_N\chi_{I_i^N}\right| \cdot \left|u_N\chi_{I_i^N}\right| \leq K\beta_N\varepsilon_N.$$

Hence

$$a^{-1}K^{-1}\beta_N^{-1}\varepsilon_N \leq \left|I_i^N\right| \cdot \left|v_N\chi_{I_i^N}\right| \cdot \left|u_N\chi_{I_i^N}\right| \leq aK\beta_N\varepsilon_N.$$

Since $\beta_N \to 1$ as $N \to \infty$, the claim follows.

We can now proceed to finish the proof of the proposition. Since by Claim 1, $\varepsilon_N \to 0$, Claim 5 yields that

$$\max\left\{\left|I_i^N\right| : i = 1, ..., N\right\} \to 0 \text{ as } N \to \infty;$$

by virtue of Claim 4,

$$\left|I_i^N\right|^{-2\delta_{N,i}} \leq e^{2\psi\left(\left|I_i^N\right|\right)} \leq e^{2\psi\left(\max\left\{\left|I_i^N\right| : i=1,...,N\right\}\right)} := \gamma_N \to 1$$

as $N \to \infty$. Put $\alpha_N = \beta_N\gamma_N$: then

$$\alpha_N \to 1 \text{ as } N \longrightarrow \infty$$

and it follows from Claim 2 that

$$\alpha_N^{-1}\varepsilon_N \leq A_{q_N(\cdot)}\left(I_i^N\right) \leq \alpha_N\varepsilon_N. \tag{5.11}$$

Moreover, Claim 5 yields

$$N\varepsilon_N = C\sum_{i=1}^{N} C^{-1}\varepsilon_N \leq C\sum_{i=1}^{N}\left|I_i^N\right| \cdot \left|v_N\chi_{I_i^N}\right| \cdot \left|u_N\chi_{I_i^N}\right|$$
$$\leq 2\|v_N\|_q\|u_N\|_{p'} \leq \infty$$

which gives, by (5.11),

$$N\varepsilon_N\left(\alpha_N^{-1}\varepsilon_N - 1\right) = \sum_{i=1}^{N}\left(\alpha_N^{-1}\varepsilon_N - \varepsilon_N\right) \leq \sum_{i=1}^{N} A_{q_N(\cdot)}\left(I_i^N\right) - N\varepsilon_N$$
$$\leq \sum_{i=1}^{N}\left(\alpha_N\varepsilon_N - \varepsilon_N\right) = N\varepsilon_N\left(\alpha_N - 1\right).$$

Thus

$$\left|\sum_{i=1}^{N} A_{q_N(\cdot)}\left(I_i^N\right) - N\varepsilon_N\right| \to 0 \text{ as } N \to \infty.$$

On the other hand we have, by Lemma 5.11 (recall again that q_N is constant on I_i^N,

$$\sum_{i=1}^{N} A_{q_N(\cdot)}^{T_{v_N,u_N}} \left(I_i^N \right)$$

$$= \frac{1}{2\pi} \sum_{i=1}^{N} |v_N| \cdot |u_N| \left(q_N'(\cdot) q_N(\cdot)^{q_N(\cdot)-1} \right)^{1/q_N(\cdot)} \sin\left(\pi/q_N(\cdot)\right) |I_i^N|$$

$$\to \frac{1}{2\pi} \int_I |v_N| \cdot |u_N| \left(p'(x)p(x)^{p(x)-1} \right)^{1/p(x)} \sin(\pi/p(x)) dx$$

as $N \to \infty$;

moreover, by definition (of v_N and u_N) one has $v_N \to v$ and $u_N \to u$ in I and so

$$\lim_{N\to\infty} N\varepsilon_N = \frac{1}{2\pi} \int_I |v||u| \left(p'(x)p(x)^{p(x)-1} \right)^{1/p(x)} \sin(\pi/p(x)) dx.$$

Since ε_N depends monotonically on N it is not difficult to see that

$$\lim_{N\to\infty} N\varepsilon_N = \lim_{\varepsilon\to0} \varepsilon N(\varepsilon),$$

consequently

$$\lim_{\varepsilon\to0} \varepsilon N(\varepsilon) = \frac{1}{2\pi} \int_I |v||u| \left(p'(x)p(x)^{p(x)-1} \right)^{1/p(x)} \sin(\pi/p(x)) dx.$$

The proof is complete. $\qquad\square$

Next, given $p \in \mathcal{SLH}(I)$, we construct approximating step-functions of p as follows. Let $N \in \mathbb{N}$ and use Lemma 5.25, applied to the function $D := D_p$: there exists $\varepsilon > 0$ such that $N_D(\varepsilon) = N$ and there are non-overlapping intervals I_i^D $(i = 1, ..., N)$ that cover I and are such that $D(I_i^D) = \varepsilon$ for $i = 1, ..., N$. Define

$$p_{D,N}^+(x) = \sum_{i=1}^{N} p_{I_i^D}^+ \chi_{I_i^D}(x), \ p_{D,N}^-(x) = \sum_{i=1}^{N} p_{I_i^D}^- \chi_{I_i^D}(x);$$

step-functions $p_{B,N}^+$ and $p_{B,N}^-$ are defined in an exactly similar way, with the function B in place of D and with intervals I_i^B arising from the use of that part of Lemma 5.25 related to B.

Lemma 5.31. *Let $p \in \mathcal{SLH}(I)$, $u,v \in C(I)$ and $N \in \mathbb{N}$. Let $\varepsilon > 0$ correspond to N in the sense of Lemma 5.25, applied to B, so that $N_B(\varepsilon) = N$, and write*

$$p^-(x) = p_{B,N}^-(x), \ p^+(x) = p_{B,N}^+(x),$$

where $p_{B,N}^+$ and $p_{B,N}^-$ are defined as indicated above. Then

$$a_{N+1} \left(T_{v,u} : L_{p^-}(I) \to L_{p^+}(I) \right) \le \varepsilon.$$

Proof.   In the notation of Lemma 5.25, there are intervals I_i^B such that $B(I_i^B) = \varepsilon$ for $i = 1, ..., N$. For each i there exists $y_i \in I_i^B$ such that

$$B(I_i^B) = \sup \left\{ \left\| v(\cdot) \int_{y_i}^{\cdot} f(x)u(x)dx \right\|_{p^+, I_i^B} : \|f\|_{p^-, I_i^B} \leq 1 \right\}.$$

Define

$$P_N f(x) = \sum_{i=1}^{N} v(y_i) \int_a^{y_i} f(y)u(y)dy \cdot \chi_{I_i^B}(x);$$

plainly P_N has rank N. Let $f \in L_{p^-(\cdot)}(I)$ and set

$$\lambda_0 = \varepsilon \|f\|_{p^-, I}. \tag{5.12}$$

Then

$$1 = \int_I \left| \frac{f(x)}{\lambda_0/\varepsilon} \right|^{p^-(x)} dx = \sum_{i=1}^{N} \int_{I_i^B} \left| \frac{f(x)}{\lambda_0/\varepsilon} \right|^{p^-(x)} dx.$$

Recall that on I_i^B the functions p^- and p^+ have constant values p_i^-, p_i^+, say, respectively, with $p_i^+/p_i^- \geq 1$. Thus

$$1 \geq \sum_{i=1}^{N} \left(\int_{I_i^B} \left| \frac{f(x)}{\lambda_0/\varepsilon} \right|^{p_i^-} dx \right)^{p_i^+/p_i^-} = \sum_{i=1}^{N} (\varepsilon/\lambda_0)^{p_i^+} \left(\int_{I_i^B} |f(x)|^{p_i^-} dx \right)^{p_i^+/p_i^-}.$$

Using the fact that

$$\varepsilon = \sup_f \left(\int_{I_i^B} \left| v(x) \int_{y_i}^x f(y)u(y)dy \right|^{p_i^+} dx \right)^{1/p_i^+} / \left(\int_{I_i^B} |f(y)|^{p_i^-} dy \right)^{1/p_i^-}$$

one obtains

$$1 \geq \sum_{i=1}^{N} (1/\lambda_0)^{p_i^+} \int_{I_i^B} \left| v(x) \int_{y_i}^x f(y)u(y)dy \right|^{p_i^+} dx$$

$$= \sum_{i=1}^{N} \int_{I_i^B} \left| \frac{v(x) \int_{y_i}^x f(y)u(y)dy}{\lambda_0} \right|^{p_i^+} dx$$

$$= \int_I \left| \frac{(T_{v,u} - P_N)(f)(x)}{\lambda_0} \right|^{p^+(x)} dx,$$

from which it follows that $\|(T_{v,u} - P_N)f\|_{p^+, I} \leq \lambda_0$. Using the definition (5.12) of λ_0 we see that

$$\|(T_{v,u} - P_N)f\|_{p^+, I} \leq \varepsilon \|f\|_{p^-, I},$$

and so $a_{N+1}\left(T_{v,u} : L_{p^-}(I) \to L_{p^+}(I)\right) \leq \varepsilon$, as claimed. $\square$

We next obtain a lower estimate for the Bernstein numbers.

Lemma 5.32. *Let $p \in \mathcal{SLH}(I)$ and $v, u \in C(I)$ and $N \in \mathbb{N}$. Let $\varepsilon > 0$ correspond to N in the sense of Lemma 5.25, applied to D, so that $N_D(\varepsilon) = N$, and write*

$$p^-(x) = p^-_{D,N}(x), \ p^+(x) = p^+_{D,N}(x),$$

where $p^+_{D,N}$ and $p^-_{D,N}$ are defined as indicated above. Then

$$b_N\left(T_{v,u} : L_{p^+}(I) \to L_{p^-(\cdot)}(I)\right) \geq \varepsilon.$$

Proof. In the notation of Lemma 5.25, there are intervals I_i^D such that $D(I_i^D) = \varepsilon$ for $i = 1, ..., N$. Since T is compact, for each i there exists $f_i \in L_{p^+(\cdot)}(I_i^D)$, with supp $f_i \subset I_i^D$, such that

$$\|T_{v,u}f_i\|_{p^-,I_i^D} \, / \, \|f_i\|_{p^+,I_i^D} = \varepsilon, \tag{5.13}$$

and $T_{v,u}f_i(c_i) = T_{v,u}f_i(c_{i+1}) = 0$, where c_i and c_{i+1} are the endpoints of I_i^D. On each I_i^D the functions p^- and p^+ are constant; denote these constant values by p_i^- and p_i^+, respectively and note that $p_i^- / p_i^+ \leq 1$. Set

$$X_N = \left\{ f = \sum_{i=1}^{N} \alpha_i f_i : \alpha_i \in \mathbb{R} \right\}.$$

Thus $\dim X_N = N$. Choose any non-zero $f \in X_N$ and put $\lambda_0 = \varepsilon \|f\|_{p^+,I}$. Then

$$1 = \int_I \left| \frac{f(x)}{\lambda_0/\varepsilon} \right|^{p^+(x)} dx = \sum_{i=1}^{N} \int_{I_i^D} \left| \frac{f(x)}{\lambda_0/\varepsilon} \right|^{p_i^+(x)} dx$$

$$\leq \sum_{i=1}^{N} \left(\int_{I_i^D} \left| \frac{f(x)}{\lambda_0/\varepsilon} \right|^{p_i^+(x)} dx \right)^{p_i^-/p_i^+}$$

$$= \sum_{i=1}^{N} (\varepsilon/\lambda_0)^{p_i^-} \left(\int_{I_i^D} |f(x)|^{p_i^+(x)} dx \right)^{p_i^-/p_i^+}$$

$$= \sum_{i=1}^{N} (\varepsilon/\lambda_0)^{p_i^-} \left(\int_{I_i^D} |\alpha_i f_i(x)|^{p_i^+} dx \right)^{p_i^-/p_i^+}.$$

An application of (5.13) shows that

$$1 \leq \sum_{i=1}^{N} (1/\lambda_0)^{p_i^-} \int_{I_i^D} |T(\alpha_i f_i)(x)|^{p_i^-} dx$$

$$= \sum_{i=1}^{N} \int_{I_i^D} \left| \frac{T_{v,u}f(x)}{\lambda_0} \right|^{p_i^-} dx = \int_I \left| \frac{T_{v,u}f(x)}{\lambda_0} \right|^{p^-(x)} dx,$$

from which it follows that

$$\varepsilon \leq b_N \left(T : L_{p^+(.)}(I) \to L_{p^-(.)}(I)\right),$$

and the proof is complete. $\qquad\square$

Theorem 5.18. *Let* $p \in \mathcal{SLH}(I)$ *and* u, v *be continuous on* I. *For all* $N \in \mathbb{N}$ *denote by* ε_N *the numbers defined by* $N = N_B(\varepsilon_N)$. *Then there are sequences* K_N, L_N, *with* $K_N \to 1$ *and* $L_N \to 1$ *as* $N \to \infty$, *such that*
(i) $a_{N+1}(T_{v,u} : L_p(I) \to L_p(I)) \leq K_N \varepsilon_N$,
(ii) $b_N(T_{v,u} : L_p(I) \to L_p(I)) \geq L_N \varepsilon_N$.

Proof. Because of the multiplicative property (S3) of the approximation numbers, $a_{N+1}(T_{v,u} : L_p(I) \to L_p(I))$ is majorised by

$$\left\| id_N^- : L_p(I) \to L_{p_{B,N}^-}(I) \right\|$$
$$\times a_{N+1}(T_{v,u} : L_{p_{B,N}^-}(I) \to L_{p_{B,N}^+}(I)) \times \left\| id_N^+ : L_{p_{B,N}^+}(I) \to L_p(I) \right\|,$$

where id_N^- and id_N^+ are the obvious embedding maps, while $p_{B,N}^+, p_{B,N}^-$ are the same as in Lemma 5.31, as is I_i^B, to be used next. Since $\left| I_i^B \right| \to 0$ when $N \to \infty$, and p is continuous, it is clear that, as $N \to \infty$,

$$\left\| p - p_{B,N}^- \right\|_{\infty,I} \to 0 \text{ and } \left\| p - p_{B,N}^+ \right\|_{\infty,I} \to 0.$$

Thus by Corollary 1.6,

$$\left\| id_N^- : L_p(I) \to L_{p_{B,N}^-}(I) \right\| \to 1 \text{ and } \left\| id_N^+ : L_{p_{B,N}^+}(I) \to L_p(I) \right\| \to 1$$

as $N \to \infty$. Part (i) now follows from Lemma 5.31. The proof of (ii) is similar, obtained via the aid of Lemma 5.32. $\qquad\square$

Theorem 5.19. *Let* $p \in \mathcal{SLH}(I)$, *and* $v, u \in C(I)$. *Then*

$$\lim_{n \to \infty} n s_n(T_{v,u} : L_p(I) \to L_p(I))$$
$$= \frac{1}{2\pi} \int_I |v||u| \left\{ p'(t) p(t)^{p(t)-1} \right\}^{1/p(t)} \sin\left(\pi/p(t)\right) dt,$$

where s_n *denotes the* n^{th} *approximation, Gelfand, Kolmogorov or Bernstein number of* T.

Proof. Use Theorem 5.18, Proposition 5.2 and the inequalities of Theorems 5.4 and 5.5. $\qquad\square$

The proofs of Theorems 5.17 and 5.19 may be combined to give the following theorem, which contains both these results.

Theorem 5.20. *Let I be representable as the finite union of non-overlapping intervals J_i $(i = 1, ..., m)$ and suppose that $p \in \mathcal{SLH}(I_i)$ and $v, u \in C(I_i)$ for each $i \in \{1, 2, ..., m\}$. Then*

$$\lim_{n \to \infty} n s_n(T_{v,u} : L_p(I) \to L_p(I))$$

$$= \frac{1}{2\pi} \int_I |v||u| \left\{ p'(t) p(t)^{p(t)-1} \right\}^{1/p(t)} \sin\left(\pi/p(t)\right) dt,$$

where s_n denotes the nth approximation, Gelfand, Kolmogorov or Bernstein number of $T_{v,u}$.

Notes

We underline the fact that the special case of Theorem 5.20 when $u = v = 1$ was given in [31] and that the proof of the above theorem follows along the same general lines as the proof of the simpler case from [31]. Also ideas from [31] were based on the techniques used for obtaining results for a Sobolev embedding on an interval.

Bibliography

[1] Acerbi, E, Fusco, N., A transmision problem in the calculus of variations, Calc. Var. 2,1-16 (1994).

[2] Acerbi, E., Mingione, G., Regularity Results for a Class of Functionals with Non-Standard Growth, Arch. Rational Mech. Anal. 156(2001) 121-140.

[3] Acerbi, E., Mingione, G., Gradient estimates for the $p(x)$-Laplacean system, J. reine angew. Math. (584)(2005), 117-148.

[4] Adams, R.A., *Sobolev Spaces*, Academic Press, New York, 1975.

[5] Alkhutov, Y., The Harnack Inequality and the Hölder Property of Solutions of Nonlinear Elliptic Equations with a Nonstandard Growth Condition. Diff. Eq. 1997,33(12)1653-1662

[6] Alkhutov, Y., Hölder continuity of $p(x)$-harmonic functions, Sbornik: Mathematics 196:2 147-71 (2005).

[7] Almeida, A. Samko, S., Embeddings of variable Hajłasz-Sobolev spaces into Hölder spaces of variable order, J. Math. Anal. Appl. Volume 353, Issue 2, 15 May 2009, 489496

[8] Anane, A., Simplicité et isolation de la premiére valeur propre du p-Laplacien avec poids, C.R. Acad. Sci. Paris, t. 305, Série I, p. 725-728, (1987)

[9] Andreianov, B., Bendahmane, M., Ouaro, S., Structral stability for variable exponent elliptic problems I: The $p(x)$-Laplacian kind problems, Nonl. Anal. 73(2010) 2-24.

[10] Appell, J., De Pascale, E. and Vignoli, A., *Nonlinear Spectral Theory*, Walter De Gruyter, Berlin, 2004.

[11] Bennett, C. and Sharpley, R., *Interpolation of operators*, Academic Press, New York, 1988.

[12] Bennewitz, C., *Approximation numbers=singular values*, J. Comp. Appl. Math. **208** (1) (2007)102-110.

[13] Binding, P., Rynne, B., Variational and non-variational eigenvalues for the p-Laplacian, J. Diff. Equations 244(2008)24-39.

[14] Cafarelli, L. Interior a priori estimates for solutions of fully nonlinear equations, Ann. Math.130 (1989) 189-213.

[15] Cafarelli, L., Peral, I., On $W^{1,p}$ estimates for elliptic equations in divergence form, Comm. Pure Appl. Math.51 (1998)1-21 189-213.

[16] Carriero, M, De Luca L., *Introduzione al Calcolo delle Variazioni*, Quaderno 1/2010, Dipartamento di Matematica E. Di Giorgi, Università del Salento, Lecce, 2010.

[17] Coscia, A., Mingione, G., Hölder continuity of the gradient of $p(x)$-harmonic mappings, C.R. Acad. Sci. Paris t328 Serie 1 p 363-368 (1999).

[18] Cruz-Uribe, D., Fiorenza, A., *Variable Lebesgue Spaces, Foundations and Harmonic Analysis*, Birkhäuser, Heidelberg-New York-Dordrecht-London, 2013.

[19] Dacorogna, B., *Direct Methods in the Calculus of Variations*, Springer-Verlag, Berlin,1989.

[20] De Giorgi, E. Sulla diferenziabilitá e l'analiticitá delle estremali degli li integrali multipli regolari, Mem. Accad. Sci. Torino Cl. Sci. Fis. Mat. Natur. (3), 3 (1957), 25-43.

[21] Deimling, K., *Nonlinear functional analysis*, Springer-Verlag, Berlin-Heidelberg-New York-Tokyo, 1985.

[22] Diening, L.,*Maximal functions on generalized $L^{p(x)}$ spaces*, Math. Inequal. Appl. **7** (2004), 245-253.

[23] Diening, L., Harjulehto, P., Hästö, P. and Růžička, M., *Lebesgue and Sobolev spaces with variable exponent*, Lecture Notes in Mathematics , Springer.

[24] Diening, L., Hästö, P., Variable exponent trace spaces, Stud. Math. 183(2)(2007)

[25] Dinca, G. and Matei, P., *Geometry of Sobolev spaces with variable exponent: smoothness and uniform convexity*, C. R. Acad. Sci. Paris Ser. I **347** (2009), 885-889.

[26] Dinca, G. and Matei, P., *Geometry of Sobolev spaces with variable exponent and a generalization of the p−Laplacian*, Analysis and Applications **7** (4) (2009), 373-390.

[27] Edmunds, D.E. and Evans, W.D., *Spectral theory and differential operators*, Oxford, Mathematical Monographs, Oxford University Press, Oxford, 1987.

[28] Edmunds, D.E. and Evans, W.D., *Hardy operators, function spaces and embeddings*, Springer, Berlin-Heidelberg-New York, 2004.

[29] Edmunds, D.E., Gurka, P. and Pick, L., *Compactness of Hardy-type integral operators in weighted Banach function spaces*. Studia Math. 109 (1994), no. 1, 7390.

[30] Edmunds, D.E., Kokilashvili, V. and Meskhi, A. *On the boundedness and compactness of weighted Hardy operators in spaces $L^{p(x)}$*. Georgian Math. J. 12 (2005), no. 1, 2744.

[31] Edmunds, D.E. and Lang, J., *Eigenvalues, embeddings and generalised trigonometric functions*, Lecture Notes in Mathematics **2016**, Springer, 2011.

[32] Edmunds, D.E., Lang, J. and Nekvinda, A., *On $L^{p(x)}$ norms*, Proc. Roy. Soc. London A**455** (1999), 219-225.

[33] Edmunds, D.E., Lang, J. and Nekvinda, A., *Some s−numbers of an integral operator of Hardy type on $L^{p(\cdot)}$ spaces*, J. Functional Anal. **257** (2009), 219-242.

[34] Edmunds, D.E. and Rákosník, J., *Density of smooth functions in $W^{k,p(x)}(\Omega)$*,

Proc. Roy. Soc. London Ser. A, 437 (1992), 229-236

[35] Edmunds, D. and Rákosník, J., Sobolev Embeddings with variable exponent, Studia Math. 143 (3)2000.

[36] Edmunds, D. E. and Tylli, H.-O., *On the entropy numbers of an operator and its adjoint*, Math. Nachr. **126** (1986), 231-239.

[37] Evans, L. C. and Gariepy, R.F., *Measure theory and fine properties of functions*, CRC Press, Boca Raton, Florida, 1992.

[38] Fabian, M., Habala, P., Santalucía, V.M., Pelant, J. and Zizler, V., *Functional analysis and infinite-dimensional geometry*, Springer-Verlag, Berlin-Heidelberg-New York, 2001.

[39] Fan, X. *Remark on eigenvalue problems involving the $p(x)$-Laplacian*, J. Math. Anal. Appl. **352**(2009), 85-98.

[40] Fan, X., *Global $C^{1,\alpha}$ regularity for variable exponent elliptic equations in divergence form*, J. Diff. Eq. 235 (2007) 397-417.

[41] Fan, X., *Eigenvalues of the $p(x)$-Laplacian Neumann problems*, Nonl. Anal. 67 (2007) 2982-2992.

[42] Fan, X. and Zhang, Q. , *Existence of solutions for $p(x)$-Laplacian Dirichlet Problem*, Nonlinear Anal. 52 (2003):1843-1852.

[43] Fan, X. and Zhang, Q., Zhao, D. , *Eigenvalues for $p(x)$-Laplacian Dirichlet Problem*, J. Math. Anal. Appl. 302 (2005):306-317.

[44] Fan, X., Zhao, D., *The quasi-minimizer of integral functionals with $m(x)$ growth conditions*, Nonlinear Anal. 39 (2000):807-816.

[45] Fan, X., Zhao, D., *Regularity of minimizers of variational integrals with continuous $p(x)$-growth conditions*, Chinese Journal of Contemporary Math 17, 4 (1996):327-336.

[46] Fan, X., Shen, J. and Zhao, D., *Sobolev Embedding Theorems for Spaces $W^{k,p(x)}(\Omega)$*, Jour. Math. Anal. Appl. **262** (2001), 749-760.

[47] Franzina, G., Lindqvist, P., , *An eigenvalue problem with variable exponents*, Nonlinear Analysis **85** (2013), 1-16.

[48] Gilbarg, D. and Trudinger, N., *Elliptic Partial Differential Operators of Second Order*, Springer-Verlag, Berlin-Heidelberg-New York, 2001.

[49] Hajłasz, P., Sobolev spaces on arbitrary metric space, Pot. Anal.5 (1996), 403-415.

[50] Harjulehto, P. and Hästö, P., *Sobolev inequalities for variable exponents attaining the values 1 and n*, Publ. Mat. 52 (2008), 347-363.

[51] Hutton, C. V., *Approximation numbers of bounded linear operators*, Dissertation, Louisiana State University, Baton Rouge, 1973.

[52] John, F., Nirenberg, L. On functions of bounded mean oscillation, Comm. Pure and Appl. Math. Volume 14, Issue 3, pages 415426, August 1961.

[53] Kenig, C., *Harmonic Analysis Techniques for Second Order Elliptic Boundary Value Problems*, CBMS-RCSM, 83, Amer. Math. Soc., Providence, RI,1994.

[54] Kokilashvili, V., Meskhi, A., *Two-weight inequalities for fractional maximal functions and singular integrals in $L^{p()}$ spaces*. Problems in mathematical analysis. No. 55. J. Math. Sci. (N. Y.) 173 (2011), no. 6, 656673.

[55] Kováčik, O. and Rákosník, J., *On spaces $L^{p(x)}(\Omega)$ and $W^{k,p(x)}(\Omega)$*,

Czechoslovak Math. J. **41** (1991), 592-618.

[56] Kufner, A., John, O. and Fučik, S., *Function spaces*, Academia, Prague, 1977.

[57] Ladyzhenskaya O. A. and Ural'tseva, N. N., *Quasilinear elliptic equations and variational problems with many independent variables* 73, 1964, pp. 172-220.

[58] Lang, J. and Méndez, O., *Modular eigenvalues of the Dirichlet p()-Laplacian and their stability*, Spectral theory, function spaces and inequalities, 125137, Oper. Theory Adv. Appl., **219**, Birkhuser/Springer Basel AG

[59] Lang, J. and Nekvinda, A., *A difference between the continuous and the absolutely continuous norms in Banach function spaces*, Czech. Math. J. **47** (1993), **221-232.**

[60] Lindenstrauss, J. and Tzafriri, L., *Classical Banach spaces I and II*, Springer-Verlag, Berlin-Heidelberg-New York, 1977 and 1979.

[61] Lindqvist, P., *Stability for the solutions of $div(|\nabla u|^{p-2}\nabla u) = f$ with varying p*, Journal Math. Anal. Appl. 127 1 (1987)93-102.

[62] Lindqvist, P., *On non-Linear Rayleigh Quotients*, Pot. Anal. 2(1993)199-218.

[63] Lions, J.-L., *Quelques méthodes de résolution des problèmes aux limites non linéaires*, Dunod, Paris, 1969.

[64] Lukeš, J., Pick, L. and Pokorný, D., *On geometric properties of the spaces $L^{p(x)}$*, Rev. Mat. Complut. **24** (2011), 115-130.

[65] Mihailescu, M , Radulescu, V. *On a nonhomogenoeus quasilinear eigenvalue problem in Sobolev spaces with variable exponent*, Proc. Amer. Math. Soc. **135**(2007), 2929-2937.

[66] Mihailescu, M , Radulescu, V. *Continuous spectrum for a class of nonhomogeneous differential operators*, Manuscripta Mathematica **125**(2008), 157-167.

[67] Mihailescu, M , Radulescu, V. *On an eigenvalue problem involving the $p(x)$-Laplace operator plus a non-local term*, Diff. Equations and Appl. Vol 1 **3**(2009), 367-378.

[68] Moser, J, On Harnack's theorem for elliptic differential equations, Comm. Pure Appl. Math. 14 (1961), 577591.

[69] McShane, E., *Extension of range of functions*, Bull. Amer. Math. Soc. 40 (1934), 837-842.

[70] Musielak, J., *Orlicz spaces and modular spaces*, Springer, Berlin, 1983.

[71] Nash, J., Continuity of solutions of parabolic and elliptic equations, Amer. J. Math. 80 (1958), 931954.

[72] Pietsch, A., *History of Banach spaces and linear operators*, Birkhäuser, Boston-Basel-Berlin, 2007.

[73] Pietsch, A., $s-$numbers of operators in Banach spaces, Studia Math. **51** (1974), 201-223.

[74] Pietsch, A., *Operator ideals*, North Holland, 1980.

[75] Pinkus, A., $n-widths$ in approximation theory, Sppringer-Verlag, Berlin, 1985.

[76] Růžička, M., *Electrorheological fluids: modeling and mathematical theory*, Lecture Notes in Math., vol. **1748**, Springer-Verlag, Berlin, 2000.

[77] Serrin, J., Local behavior of solutions of quasi-linear equations, Acta Mathematica 111: 247302.

[78] Taylor, A.E., *Introduction to functional analysis, Wiley, New York, 1958.*

[79] Yao, J., *Solutions for Neumann boundary value problems involving $p(x)$-Laplace operators,* Nonl. Anal. 68 (2008), 1271-1283.

[80] Zeidler, E., *Nonlinear Functional Analysis and its Applications III, Variational Methods and Optimization,* Springer Verlag, New York, 1985.

Author Index

Subject Index

Notation Index